Joseph Parry

Water

Its Composition, Collection and Distribution - A Practical Handbook for Domestic and General Use

Joseph Parry

Water
Its Composition, Collection and Distribution - A Practical Handbook for Domestic and General Use

ISBN/EAN: 9783337139346

Printed in Europe, USA, Canada, Australia, Japan

Cover: Foto ©berggeist007 / pixelio.de

WATER:

ITS

COMPOSITION, COLLECTION,

AND

DISTRIBUTION.

A PRACTICAL HANDBOOK FOR DOMESTIC AND GENERAL USE.

By JOSEPH PARRY, C.E.

WITH ILLUSTRATIONS.

LONDON:
FREDERICK WARNE AND CO.,
BEDFORD STREET, STRAND.

DALZIEL BROTHERS, CAMDEN PRESS, LONDON, N.W

PREFACE.

We live in an age of sanitary restlessness. Never has there been so much solicitude about drains and dust, about the air we breathe and the Water we drink. Never have there been so many sanitary schoolmasters in the world.

But, in the practical application even of the knowledge we already possess, the progress that has been made is exceedingly small, when compared with the magnitude of the work to be done. To effect greater progress in the future, there must be a wider diffusion of knowledge concerning the laws of health, and a more general recognition of individual responsibility in relation to them.

Water is an indispensable agent in every kind of sanitary improvement, and the adoption of a high standard of purity cannot fail to exercise an ameliorating influence. The subject of Water Supply has therefore a paramount claim to attention, and it derives additional importance from its intimate connection with such questions as the purification of rivers, the prevention and utilization of floods, and the promotion of temperance.

This book has been written with a desire to give information that may be useful to the public in reference to Water and Water Supply; in deciding upon the merits of Water schemes and methods of distribution; and in the selection, construction, and management of the numerous Water appliances in which consumers are interested. In the part relating to house fittings, I have endeavoured to take up those points in regard to which, according to the experience of every waterworks engineer, information is most frequently sought and required. The employment of technical phraseology has, as far as possible, been avoided.

J. P.

LIVERPOOL,
January 1, 1881.

CONTENTS.

WATER.

CHAPTER I.

Water, its use and abuse—Domestic supply—Health of towns inquiry, 1844—Subsequent progress—Review of water supply legislation—Pollution Commissioners' Report—Outbreaks of disease due to impure water—Parliamentary Return, 1879—Present state of water supply in England—What remains to be done—General and local regulations—How further progress is to be effected.

ABOUT three feet of water, in the form of rain, snow, and hail, falls upon the surface of Great Britain every year. Spread over the present estimated population, the average daily fall is more than nineteen tons to every inhabitant.

This represents a vast power for good: to cleanse and refresh the air and earth; to irrigate and fertilize the soil; to support life; to drive machinery; to give wholesome drink to man and beast. Uncontrolled, it sweeps over valleys and lowlands in destructive floods. Misapplied, it becomes a carrier of disease and death.

To preserve and convey, for the use of every town and village and of every household throughout the country, a sufficient supply of the water which nature has so abundantly provided, is a service of inestimable value and importance. There is no sign more convincing and satisfactory of the progress which the country has made during the second half of the present century than is afforded by the numerous and costly works that have been carried out for the collection and distribution of water for domestic purposes.

In the year 1844 a Royal Commission was issued to inquire into the state of large towns and populous districts in England and Wales. The instructions included "the supply of water in such towns and districts, whether for purposes of health or for the better protection of property from fire." Out of the fifty towns to which the inquiries of the Commissioners extended, there were only six in which the arrangements and supplies could be deemed in any comprehensive sense good, while in thirteen they were indifferent, and in thirty-one bad. There was almost universal scarcity of supplies for domestic use among the poorer classes. In some of the larger towns the proportion of houses receiving a separate supply was extremely small. The waterworks were generally in the hands of joint-stock companies, who, being trading bodies, only laid their pipes to those parts where they got the largest and best customers. The mode of supplying and charging operated most prejudicially to the interests of the poor. Where pipes were laid, the water was usually distributed on a restricted intermittent system, in many places a supply of only one or two hours' duration, two or three times in a week, being given. There was no general law applicable to the subject, and it was not generally recognized as part of the duty of the body entrusted with the local government of a town to enforce or to provide an ample supply of water.

The Commissioners recommended that it should be rendered imperative on local administrative bodies to procure sufficient supplies of water, for all domestic, public, and sanitary purposes; that water companies should be required to comply with the demands of local administrative bodies on equitable terms; that local administrative bodies should be empowered to purchase waterworks by agreement; that new companies should only be established on condition that the local administrative body might subsequently purchase the undertaking; that where pipes were laid down all houses capable of benefiting by them should be rated in the same way as for sewerage and other local purposes, the owners of small tenements being made liable for the rates; that for increasing the protection of property from fire, there should be a constant supply at high pressure.

I have given this short summary of the conclusions

and recommendations of the celebrated Health of Towns Commission, both for its historical value, and because it provides a convenient point from which to measure the progress that has since been made.

Forty years ago water was sold in the streets of large towns at a halfpenny or penny the bucket, from casks mounted on wheels. To-day there is no town of importance without public works from which water is distributed through pipes to every dwelling.

In 1817 the first general Act relating to water supply was passed. The principle was laid down that the body undertaking to supply a town or district should provide pure and wholesome water sufficient for the domestic use of all the inhabitants, and constantly laid on at a pressure adequate to reach the top storey of the highest house. The undertakers were required to lay pipes to any houses in their district of which the aggregate water rates payable annually should be not less than one-tenth the cost of extending the pipes if the owners or occupiers agreed to pay such rates for three years. It was also provided that the undertakers should, on the request of Town Commissioners, affix fire-plugs to their mains.

The Report of the Health of Towns Commission

led to the passing of the Public Health Act, 1848. The preamble set forth that "It is expedient that the supply of water to such towns and places, and the sewerage, drainage, cleansing, and paving thereof, should as far as practicable be placed under one and the same local management and control." With this view, local authorities were empowered to provide sufficient supplies of water by contract or agreement, and to erect waterworks, subject to a proviso that, where a waterworks company already existed, new works could only be constructed if the company after demand failed to give a proper supply on reasonable terms. Waterworks companies were empowered to contract with local authorities for the supply of water, or to sell or lease their waterworks to local authorities willing to take them. It was also enacted that a local authority might require the occupier of any house to obtain a proper supply of water, where such supply could be furnished at a rate not exceeding twopence per week.

Further powers and facilities were given to local authorities for constructing and purchasing waterworks under the Local Government Act, 1858; Amendment Act, 1861; and the Sanitary Acts of 1866 and 1874.

An Act was passed in 1870 (the Gas and Water Facilities Act) to facilitate in certain cases the obtaining of powers for the construction of waterworks, by provisional orders to be granted by the Board of Trade, after notice and inquiry. Provision was also made for small water companies not incorporated under a special Act, to obtain the necessary parliamentary powers if they got the consent of the local authority. The application of provisional orders was further extended by the Gas and Water Works Facilities Amendment Act, 1873. The Public Health Act, 1875, repealed and consolidated all previous Acts relating to sanitary matters. By the Health Act of 1872 England was divided into Urban and Rural Sanitary Districts, and an authority was thus established in every part of the country, upon which devolved the duty of securing a proper and sufficient supply of water for all domestic and sanitary purposes. The provisions of the Health Acts enable this duty to be discharged.

If a district sanitary authority fails to provide a supply of water, so that the health of the inhabitants is endangered, the Local Government Board is directed, on complaint being made, to institute an inquiry, and if satisfied that the local authority has been guilty of the alleged default, to issue an order requiring, and failing compliance compelling, the performance within a specified time of the duty that has been neglected.

In 1877 an Act was passed to give further facilities to landowners of limited interests, to charge their estates with the expense of constructing or contributing to the construction of reservoirs for the better supply of villages and towns, where it could be shown to the satisfaction of the Enclosure Commissioners that such works would be permanently productive of profit to the estates.

By the Public Health (Water) Act, 1878, it has become the duty of every rural sanitary authority to see that every occupied dwelling house within their district has, within a reasonable distance, a sufficient and available supply of wholesome water. Under the same Act it is the duty of every rural authority to make a periodical inspection of the water supply within their district, and it is enacted that no house hereafter erected or rebuilt shall be occupied until an official certificate has been given that a sufficient supply of wholesome water has been provided for the house.

This brief outline of general legislation affecting water supply would be incomplete without some reference to the private bill legislation of the same period.

In the summer of 1871 I had occasion to obtain an abstract of waterworks legislation, and I found that between 1847 and 1871 there had been 533 Water Acts passed. Of these, 310 had been obtained by companies, 209 by local authorities, and 14 were general. Of the Acts obtained by companies 124 were to incorporate new companies, 25 to re-incorporate companies, 5[illegible] contained provisions for extending the limits of supply, 116 for additional works, and 21 for the transfer of works to local authorities. Of the Acts obtained by local authorities, 47 were for the purchase of works from companies, and 56 to construct new works, 61 to provide for additional works, 26 to extend the limits of supply, 101 were limited in operation to the districts within which the authorities applying exercised jurisdiction, and 108 included places beyond those districts.

Since the year 1870 Parliament has passed 234 private Acts relating to water supply. 140 of these were obtained by local authorities, and 94 by water companies. During the same period 80 provisional orders have been granted to water companies by the Board of Trade under the Gas and Water Facilities Acts.

An elaborate Report on the Domestic Water Supply of Great Britain was issued in 1874 by the Rivers Pollution Commissioners. They examined more than two thousand samples of drinkable water; they inspected a large number of waterworks; and they applied for information by circular to all parts of the country. Their applications do not appear to have been always treated with the respect due to communications from Royal Commissioners, for in their report they say, "There are at least 16,000 cities, towns, villages, and hamlets, in

England, Scotland, and Wales; but the number we have visited or from which we, after repeated applications, have received information, is only 610." These 610 included nearly all the principal towns, and the Commissioners remark, "Every variety is copiously represented." On these data they based the following conclusions:—

1. A large majority of the cities and large towns, other than London, are abundantly supplied with palatable and wholesome water.

2. In other towns and villages and other inhabited places the water available for domestic purposes is frequently neither abundant nor wholesome.

3. Immense numbers of the people are daily exposed to the risk of infection from typhoidal discharges, and periodically to that from cholera dejections.

In regard to rural supplies they say, "The remaining twelve millions of country population derive their water almost exclusively from shallow wells, and these are, so far as our experience extends, almost always horribly polluted by sewage and by animal matters of the most disgusting origin."

The Annual Reports of the Medical Officer to the Local Government Board contain important evidence concerning the sanitary state of districts where special outbreaks of disease occur. In the last report issued, there is an abstract of fourteen medical inspections made in regard to the incidence of disease in particular places: out of the fourteen there were ten in which the inspectors reported an impure or deficient water supply. A report issued in 1874 contains 144 illustrations from inspectors, reports during four years, 1870-3, of the circumstances in which enteric fever is commonly found prevalent. In 134 out of the 144 cases examined, the inspectors reported an impure or defective water supply.

In an able Report on Filth Diseases (1874), Mr. Simon, the late Medical Officer to the Privy Council and Local Government Board, stated that he believed it to be the common conviction of persons who had studied the subject, "That the deaths which we in each year register in this country are fully 125,000 more numerous than they would be if existing knowledge of the chief cause of diseases, as affecting masses of the population, were reasonably well applied."

There is unfortunately plenty of filth, and plenty of disease, in places where the water supply is unexceptionable in quality and unstinted in quantity. A copious supply of pure water is only one of many conditions conducive to health and comfort; but it is an important condition; and the existence of evils which have a greater influence in producing a high rate of mortality, such as poverty, overcrowding, and intemperance, ought not to lessen our exertions to provide that which all experience shows to be essential to healthful existence.

The most complete information that has recently been collected, with regard to the supply of towns, is contained in a parliamentary return issued last year, "showing the means by which drinkable water is supplied to every urban sanitary district in England and Wales, such means being provided by public or private arrangements." Unfortunately some of the questions put to the urban sanitary authorities were exceedingly vague and indefinite, and were not framed so as to elicit information on matters of primary interest. The return is therefore less valuable than it might have been. Then there is no abstract given so as to reduce the information to a convenient form for judging of the general state of water supply in the country. To make up to some extent for these deficiencies, I have had the annexed table prepared, based on the return, with a view more particularly to show, in the first place, to what extent the supply of the country is now in the hands of joint-stock companies or of local administrative bodies; and in the second place, how far the constant system of distribution prevails. The return does not contain specific questions and answers on the first of these points; and the question relating to the second has evidently been misunderstood by many of the sanitary authorities.

ABSTRACT OF STATISTICS FROM "URBAN WATER SUPPLY" RETURN, 1879.

NOTE.—The Metropolis is not included, as it is not an Urban Sanitary District.

SOURCE OF SUPPLY.	Supply in hands of Local Authorities. Population in 1871.		Supply in hands of Water Companies. Population in 1871.		Part Public and Private Wells, and part Public Waterworks.	Private and Public Wells. No Public Waterworks.	Aggregate of Population supplied from the several Sources.
	Service constant.	Service intermittent.	Service constant.	Service intermittent.	Population in 1871.	Population in 1871.	
Wells	755,453	406,754	733,319	225,701	407,065	812,445	1,220,257
Gravitation	2,860,968	91,264	521,361	226,994			2,121,230
River	820,773	115,294	274,919	276,648			3,700,527
Springs	203,950	191,300	108,033	209,451			1,487,634
Gravitation and Wells	621,521	13,704	29,477	—			712,734
River and Wells	87,446	7,509	51,617	53,214			664,702
Wells and Springs	63,457	54,545	151,863	73,329			199,177
Gravitation and Springs	59,991	4,900	201,618	6,323			343,194
River and Springs	22,759	54,676	3,586	—			272,832
Gravitation, River, and Wells	426,576	6,276	9,003	30,001			81,021
Source not known	—	—	17,316	18,352			471,856
							35,668
	5,922,834	946,213	2,101,509	1,120,919	407,065	812,445	11,310,832
Number of Places	313	79	163	88	65	230	938

The population of urban districts included in the return was, according to the census of 1871, 11,310,832. This is exclusive of the metropolis, which is not an urban sanitary district. The population of the metropolis in 1871 was 3,254,260. Adding this to the figures given above, we have—

Population supplied with water by joint-stock companies	6,475,783
Supplied by local administrative bodies	6,869,017

Classifying the population according to methods of water distribution, the result is—

Under the constant supply system	8,024,343
Under the intermittent supply system	5,320,492

To arrive at the present population of these districts the increase since 1871, probably 15 per cent., must be added. Relatively, the figures cannot be far wrong.

Omitting the metropolis, it therefore appears that about 61 per cent. of the population in urban sanitary districts is supplied by local authorities, and about 28 per cent. by water companies. To 71 per cent. of the urban population water is constantly laid on, and to 18 per cent. only an intermittent supply is given.

The effect of public control is seen in the fact that constant service is given to 86 per cent. of the inhabitants of districts in which the works are managed by local authorities, against 65 per cent. (or, including the metropolis, 32 per cent.) in districts where the water-works are managed by companies.

Owing to the imperfect answers furnished by many or the sanitary authorities, it is impossible to ascertain from the return the precise number of persons residing in urban sanitary districts who are without a proper supply of water, and who are dependent on shallow wells or similar unsatisfactory sources. The abstract shows 230 districts and a population of 812,115 without any public waterworks, and 65 districts containing a population of 407,065 only partially supplied from public works. In addition to these there is a considerable number of people residing in towns and districts where works are established who do not take a supply from them.

Let us now briefly review what has been done in England for supplying the people with water.

There is no important town without waterworks from which all the inhabitants can obtain a supply, though in some towns it is deficient in quantity and inferior in quality. Nearly all the principal towns (the metropolis always excepted) are liberally supplied with wholesome water at high pressure and at low rates. In most of them the works are under the control of the local authorities. Liverpool and Manchester, already possessed of very large works, have recently obtained power to construct additional works exceeding in magnitude anything hitherto attempted in this country. In many places where water is derived from suitable sources the system of distribution is defective and wasteful, and the charges are high. The rural population is almost entirely supplied from streams and shallow wells, which are often seriously polluted.

Such being the broad facts in relation to this matter, what remains to be done to secure a wholesome supply of water for the use of every household? Is further legislation needed? I think not. Parliament may render useful service by cheapening and simplifying private bill legislation, and by encouraging joint water schemes. Beyond this, I think we have had legislation enough. What is wanted is rather to put existing laws and existing knowledge into practice. Nor is it desirable, where it can possibly be avoided, to invite the interference of a Government department. The Local Government Board has now great power, for good or evil, over the sanitary condition of the country. Under this central Board are the urban and rural district authorities, with their medical officers, armed with the comprehensive and rigorous provisions of the Public Health Acts. The difficulty is to induce district authorities to take an intelligent interest in sanitary matters, and to exercise the powers they possess.

For the removal of this difficulty very much depends on the efficiency of the medical officers. Where they are vigilant and qualified, valuable work will be done. On the other hand, where they are indolent and incapable, the local authority will probably remain inactive. In such cases the interference of the Local Government Board may be necessary. But it should be the aim of all district authorities, and of all who desire reform, to render unnecessary, and to discountenance, the interposition of the London Board.

The present tendency to centralization requires watching and guarding with great jealousy. Seeing the local apathy and active opposition by which the introduction of sanitary improvements has often been thwarted, it is not surprising that zealous reformers should urge Parliament to enforce the adoption of such measures through the agency of a Government department. But compliance may be purchased at too great a cost. Anything that tends to lessen the sense of responsibility and to repress the exercise of intelligence on the part of individual citizens and of local administrative bodies, is antagonistic to the permanent interests of the country, and opposed to the spirit of our institutions.

A Government department can seldom keep pace with the progress of scientific knowledge. Government regulations cannot be easily changed. If they are revised to suit every new discovery, people will justly complain of the expense and uncertainty to which they are subjected in consequence of frequent alterations. On the other hand, a Government is placed in a somewhat ridiculous position if its rules will not permit the use of improvements which are adopted everywhere outside of its influence.

Where Government has to interfere in local administration, its function is perhaps best discharged by laying down general principles, leaving a large amount of liberty as to the methods to be pursued.

The adoption and enforcement of sanitary and other regulations by local administrative bodies within their own limits stands clearly upon a different footing. One of the first duties of a district sanitary and water authority is to introduce proper regulations, and to see that those regulations are faithfully observed. This applies with especial force to the erection of new buildings, and is as necessary for the protection of honest tradesmen as for the general benefit of the inhabitants. There has been scarcely any effective control over house plumbing until recently. In most places it has been left very much to the discretion of builders and plumbers; and the evil effects of employing incompetent, unskilful, and unscrupulous men, are to be seen everywhere, in wasteful water fittings and defective drainage. Model regulations, which have stood the test of experience, are now furnished by the Local Government Board to all district sanitary authorities, who should be careful to employ qualified men to carry them out. How far such regulations should be made retrospective in their operation must depend on the circumstances of each case. In applying new regulations to old buildings, every possible consideration should be shown to avoid inflicting unnecessary hardships upon the owners or occupiers of property interfered with.

In this matter of neglected house plumbing, I must admit that engineers have not been free from blame. As historians have confined their narratives too much to palaces and battle-fields, so have engineers bestowed attention too exclusively upon reservoirs and aqueducts, to the neglect of domestic appliances, upon which, after all, the success of their operations depends quite as much as upon the larger works. It is of no use to build substantial sewers if the drain-pipes are so badly laid that sewage escapes under the houses. Nor is it of much avail to provide wholesome water in the mains if cisterns are allowed to be receptacles for foul gases.

For the remedy of the evils to which I have adverted, and for the general introduction and maintenance of a wholesome supply of water, I think we must look mainly to the spread of knowledge on the subject, and the formation of an enlightened public opinion.

Ignorance and apathy are the great enemies of all reform: — ignorance of evil effects, and the means of averting them; apathy, from a sense of personal helplessness, and from physical and moral degradation.

That the quality of a water supply exercises a potent influence upon health is universally admitted, but the admission does little more than reflect a popular impression, to which practical effect is seldom given. For instance, in deciding upon a change of residence, from town to country, from one town to another, or from one part of the metropolis to another, how rarely is a moment's consideration given to the difference there may be in the character of the water supply, though it may be the most serious circumstance as affecting health connected with the change.

Starting with a conviction of the vital importance of a good water supply, not only as an essential part of the animal economy, but for various purposes of cleanliness and comfort, the subject has for each individual a practical bearing, and assigns to each a personal duty. Few houses are free from sanitary defects, capable of being easily remedied. When an outbreak of disease occurs, and is traced to these defects, no exertion is spared to get rid of the danger. Is not prevention better than cure? If the remedy lies with a parsimonious landlord who will not do what is wanted, the local sanitary authority can compel compliance. If the local authority is at fault, public opinion or the Local Government Board can set it right. But the occupier is also under obligations, the discharge of which he ought not to shirk.

In connection with the modern system of separate house supplies, there are many useful and ingenious, but often complicated, sanitary appliances, which are to a considerable extent under the care and control of occupiers. The introduction of these appliances has been much more rapid and general than the education of the people in the use of them; and the ignorance that prevails causes much loss of money, comfort, and health.

People are seldom concerned about their pipes and traps, cisterns and taps, so long as they get water free from taste and colour. They trust to the waterworks authorities to exercise whatever supervision may be necessary. They overlook the fact that waterworks regulations and inspections are mainly designed to prevent waste, and to secure payment for the commodity supplied; and notwithstanding the periodical visits of waste water inspectors, if the consumer's intelligent co-operation is withheld, a great amount of waste may take place, and a supply which in the street main is wholesome, may, through imperfect domestic apparatus, be unfit for use.

In districts where no general supply is provided, it is highly necessary that knowledge should be diffused as to the precautions to be adopted, and the means to be taken, to choose the safest source available, and to insure freedom from the dangers to which those who drink polluted water are exposed. There is no rustic comsumer who cannot derive some assistance from the possession of such knowledge.

I wish to fix upon every individual a sense of personal responsibilty in these matters. It is not enough to admit in a general way their importance to the moral and physical well-being; there is something to be done, probably under your own roof, certainly within the reach of your influence.

CHAPTER II.

COMPOSITION OF WATER AND SOURCES OF SUPPLY.

Pure water, what it is—Gases in water—Natural water—Safe sources of supply—Lakes, rivers, deep and shallow wells—Sewage pollution—Rain-water—Water analysis—Organic matter in water — Standards of purity — Controversies of chemists — Action of Parliament—Effect of water supply on mortality—Diseases communicated through water—Microscopical examinations—Physical tests—Hardness of water—Softening processes.

WATER is a compound of hydrogen and oxygen, in the proportions of two atoms of hydrogen to one atom of oxygen.

It is a most powerful solvent, and it greedily absorbs, and holds in solution, the gases which are present in the atmosphere. A sample of rain-water, examined by the Rivers Pollution Commission, contained, in 100 cubic inches, 1·308 of nitrogen, 0·637 of oxygen, and 0·128 of carbonic acid; making altogether 2·073 cubic inches of atmospheric gases.

Absolutely pure water is not to be found in a natural state, either on the face or in the crust of the earth. In descending from the clouds, it carries down impurities from the air through which it passes. In flowing over or penetrating the surface of the earth, it gathers in its course foreign matters of mineral, vegetable, and animal origin. A pure natural water is clear, transparent, and, when viewed in a small volume, colourless, free from taste or smell; viewed in bulk, or in a long glass tube, it has a blue-green tint. The most wholesome waters for potable purposes are those obtained:—

1. From rivers and lakes in barren and uninhabited mountainous districts, where the rainfall is heavy and flows rapidly off the land.

2. From deep wells or springs.

The geological formations most favourable for well-sinking are: the new red sandstone, chalk, and oolites. They form excellent filters, and, as the water percolates slowly through the interstices of the rock, organic impurities are to a great extent removed, or are so changed as to become entirely innocuous. At the same time, mineral constituents are dissolved, which, when present in large quantities, render the water undesirable as a beverage, and inferior to soft water for manufacturing and culinary purposes.

Deep wells should not be sunk in populous districts, nor in proximity to cesspools. Where objectionable surroundings cannot be avoided, special precautions should be taken to prevent contamination, and the composition of the water should be frequently ascertained and closely watched.

The formations which are most valuable for their water-bearing qualities are unfortunately also the most liable to pollution. They are not homogeneous, but are largely broken by faults and fissures which may become channels for the carriage of sewage. To take as an illustration the new red sandstone, one of the best water yielding rocks in the country: suburban and rural districts on this formation have generally no system of sewers, as the sandstone is such an excellent absorbent that the sewage can be quickly got rid of by means of cesspools, the contents of which often disappear with marvellous rapidity.

Small cesspools in the sandstone often receive the drainage of houses for several years, without being emptied and without overflowing. When cesspools become clogged, their usefulness is sometimes restored by firing a charge of blasting-powder, to loosen the bottom. A deep well sunk for the supply of a town, on a site carefully selected by an experienced engineer, was surrounded by fifty-one cesspools, some of them of unusual depth, within a radius of half a mile, and many of them evidently in close sympathy with the well. After the well had been completed, the water was analysed, and found to be contaminated by organic matter to such an extent, that pumping operations had to be suspended until the district was sewered. A deep well may affect and exhaust all the neighbouring shallow wells for a distance of two or three miles, and evidence has been afforded by the action of the pumps, that the communication between such extreme points may be very rapid.

Under circumstances of this kind, there can be no doubt that noxious organic bodies may be conveyed almost as freely, and with as little change, as in a river; while the difficulty of detection is far greater. These remarks are made, not in any way to depreciate well supplies, but, simply to show the necessity of care and judgment in selecting sites for sinking, and in adopting timely measures to prevent pollution.

It cannot be too widely known or too emphatically proclaimed, that the wholesomeness of water cannot be determined by its appearance: bright sparkling water, which is so highly prized, often owes its seductive appearance and pleasant taste to contact with sewage. Shallow wells are generally polluted to a perilous extent; and persons who are dependent on such means of supply should very critically examine the surroundings, to see that there is no cesspool, drain, or other source of contamination, from which danger is possible. Rivers and streams polluted by sewage are not safe sources of supply, though there may be no visible evidence of the presence of sewage at the point where the water is abstracted.

Rain-water collected from the roofs of buildings varies in composition, according to the state of the atmosphere through which it descends, of the surface upon which it falls, and of the receptacle in which it is stored. The Rivers Pollution Commissioners obtained a large number of samples of rain-water for experimental purposes from a leaden rain-collector erected in a field at Rothampstead, twenty-five miles out of London; and also several samples of stored rain-water, collected in tanks, in various parts of the country, for domestic supplies. The Commissioners found that the samples from Rothampstead were by no means so uniformly free from impurities as it is commonly supposed that rain-water would be under such conditions. Out of the eight samples of stored rain-water they found only one fit for domestic use. The conclusion at which they arrived was that rain-water collected from the roofs of houses and stored in underground tanks is "often polluted to a dangerous extent by excrementitious matters, and is rarely of sufficiently good quality to be employed for domestic purposes with safety." They were of opinion that in Great Britain, and especially in England, we shall "look in vain to the atmosphere for a supply of water pure enough for dietetic purposes."

It must, however, not be forgotten that rain is the ultimate source of all water supplies, and if proper precautions are taken to insure the cleanliness and fitness of the receiving surface and tank, there is no reason why rain-water should not be obtained from the roofs of buildings in rural districts, as well adapted for domestic purposes as most of the water collected into reservoirs from upland watersheds. In towns and suburban districts the impure condition of the atmosphere, and the unavoidable accumulation of filth on housetops, preclude the collection of rainfall fit for dietetic use.

If the quality of a water supply is doubtful, a sample should be sent to a competent analytical chemist for examination. A qualitative analysis gives the constituent elements of the sample without the quantities. A quantitative analysis gives the proportional quantity of each of the elements contained in the sample. For a complete analysis not less than 2 gallons of water should be provided. The sample should be taken in a glass bottle, with a ground glass stopper. If a cork is used it should be new, and washed before insertion. Care should be taken that the bottle is perfectly clean. It should be filled to within about half an inch of the stopper, and immediately before being filled it should be washed out two or three times with water from the source to be sampled. In filling a bottle from a reservoir or river, the vessel should be immersed so that the top will be below the surface. The stopper should be securely tied and sealed. A statement should be sent to the chemist describing fully the source of the sample, and the circumstances under which it was taken.

To make a chemical analysis of water intelligible to those who have not studied chemistry is no part of my present purpose, but a few particulars relating to water analysis may be useful to enable an opinion to be formed with regard to the meaning and value of the most important of the figures usually supplied by analytical chemists.

The first point to be observed in reading an analysis is whether the figures refer to grains per gallon or to parts per 100,000. It was formerly the invariable practice, as it is still the practice of the majority of chemists, to express the results in grains per gallon. Many chemists now state the results in parts per 100,000. To convert grains per gallon into parts per 100,000, move the decimal point one figure to the right and divide by 7, thus:—

Suppose the grains per gallon to be 17·5—

Then 7)17·5
Equal to 25·0 parts per 100,000.

To convert parts per 100,000 into grains per gallon the process is of course the reverse, thus:—

Parts per 100,000 =	25·0
multiplied by	7
Equal to grains per gallon	17·5

The quantities given always apply to matters in solution, and not to matters in suspension, unless the latter are expressly mentioned.

The total solid matters, or total solid impurity, which is generally the first quantity given in a water analysis, represents the residue left when a sample of water has been evaporated to dryness. This varies greatly according to the source whence the water is derived. The best river-waters contain from 2 to 4 parts per 100,000. The best deep well-waters from the red sandstone contain from 14·0 to 20·0 parts per 100,000, and the chalk from 25·0 to 35·0 parts per 100,000. Water from similar sources containing a much higher total of solid impurities may be quite wholesome, but when the excess is considerable it must be regarded with suspicion. The late Dr. Parkes was of opinion that in pure and wholesome water the total solids should not exceed 8 grains per gallon, unless it were chalk-water, in which case the total solids should not exceed 14 grains per gallon of calcium carbonate. In "usable" water, he thought that the total solids should not exceed 30 grains per gallon, unless they were chiefly a mixture of sodium chloride and carbonate, in which case they might run up to 50 grains or even more without apparent bad effects.

The most important determinations of the chemist are those which relate to organic matters, but there is no method known by which the actual quantity or condition of dissolved organic matter can be accurately ascertained.

Many methods have been tried and abandoned. At present there are three processes employed which give results more or less trustworthy, and with regard to which there has been much controversy among chemists. I shall not attempt to describe these rival processes, but simply explain how the results obtained are stated.

1. *The Ammonia process.*—Where this is adopted the results are given under the heads of "Albumenoid Ammonia," "Free Ammonia," and "Chlorine," in parts per million.* Professor Wanklyn (the inventor of this method) classifies waters according to the quantity of albumenoid ammonia they yield, thus:—

	Parts per million of Albumenoid Ammonia.
(*a*) Waters of extraordinary organic purity	0·00 to 0·05
(*b*) Safe waters	0·05 to 0·10
(*c*) Dirty waters	Excdg. 0·10

* To reduce to parts per 100,000, move the decimal point one to the left.

2. *The Oxygen process.*—This is a method for determining the oxygen required to oxidize the organic matter. It was employed by the late Dr. Letheby, and since his death has been employed by Dr. Tidy, in the monthly examinations of London waters for the Society of Medical Officers of Health. Dr. Tidy divides water into four classes :—*

	Oxygen required to oxidize oxidizable matters, in parts per 100,000.
1. Waters of great organic purity	not excdg. 0·05
2. Waters of medium purity	0·05 to 0·15
3. Waters of doubtful purity	0·15 to 0·21
4. Impure waters	exceeding 0·21

This classification has been proposed as a "more accurate means of testing results than mere general conformity." Dr. Tidy has insisted emphatically that "all classifications founded on one single factor in the analysis of a water are to be accepted with great caution," and that chemists should "under no circumstances decide the value of a water from an incomplete analysis."

3. *The Combustion process of Drs. Frankland and Armstrong.* The distinctive feature of this process is

* See "Journal of Chemical Society," January, 1879, also for May, 1880.

the determination of the organic carbon and nitrogen, the chief elements of organic matter. The weight of organic carbon present is taken as an indication of the amount of organic matter with which the water is contaminated; and the origin of the organic matter, whether animal or vegetable, is inferred from the relative proportions in which the two elements are found. The Rivers Pollution Commissioners, in their sixth Report, give the following classification, based upon this process:—

"Surface-water or river-water which contains in 100,000 parts more than 0·2 part of organic carbon, or 0·3 part of organic nitrogen, is not desirable for domestic supply, and ought, whenever practicable, to be rejected.

"Spring and deep well-water ought not to contain in 100,000 parts more than 0·1 part of organic carbon, or 0·3 part of organic nitrogen. If the organic carbon reaches 0·15 part in 100,000 parts, water ought to be used only when a better supply is unattainable."

Professor Frankland has since proposed the following classification, at the same time deprecating hard and fast divisions, and reliance on partial analyses:—*

* See "Chemical News," February 14th, 1879.

	Proportion of Organic Carbon in parts per 100,000.	
	Upland Surface Water.	Other than Upland Surface Water.
Water of great organic purity	not excdg. 0·2	not excdg. 0·1
Do., medium purity	from 0·2 to 0·4	from 0·1 to 0·2
Do., doubtful purity	" 0·4 to 0·6	" 0·2 to 0·4
Impure water	more than 0·6	more than 0·4

An opinion formed with respect to the quality of water from the determination of one only of its constituents may prove to be entirely erroneous, and unless the evidence of organic pollution is sufficiently conclusive to condemn the water absolutely, a full quantitative analysis of all its parts, organic and inorganic, should be obtained. The importance of this, and at the same time the necessity of exercising care and judgment in selecting samples, are shown by the following example of one of the finest and purest mountain streams in the country, recently analysed, under various conditions of flood and drought, by Professor Frankland, for a northern town.

		Organic Carbon in parts per 100,000.
1878.		
Oct. 10th.	Stream in flood....................................	0·759
" 12th.	Flood subsided, but rate of flow still high	0·282
1879.		
March 5th.	Flood equal to 10th Oct., 1878...............	0·377
" 28th.	Stream low	0·087

3

It will be observed that by taking Professor Frankland's standard, without regard to other considerations, this most excellent water would on the 10th of October have been condemned as "impure," on the 12th would have been classified as of "medium purity," and on the 28th of March as of "great organic purity."

There is considerable diversity of opinion among chemists touching the value and significance of the several methods of analysis to which I have referred. They have all been shown to be more or less defective, and to give conflicting results. Some of the methods largely employed by chemists, and relied upon to determine the quality of a water supply, give results which are altogether misleading and untrustworthy. We are told by one school of chemists that we may unhesitatingly drink water which is condemned by another school as being unfit for human consumption, and the conclusion seems inevitable that in the present unsettled and unsatisfactory state of knowledge on the subject the guidance of chemical analysis cannot be unreservedly followed.

One of the chief points in dispute is whether a river into which sewage has been put can with safety be used for a domestic supply after flowing a given distance. The question at issue is to chemists not merely of scientific and sanitary interest; it has also an important financial aspect which gives zest to the controversy. There are waterworks authorities and manufacturers who are concerned in river supplies, and who require defence against attempts to enforce high standards of purity; and there are on the other side some water purists who push their views further than the teachings of science and experience at present warrant.

So far as the action of Parliament indicates the state of public feeling, it would seem that the tendency is decidedly in favour of raising the standard of purity, to the exclusion for domestic supplies of rivers that have been polluted, even remotely, by sewage.

In the session of 1878 there were two private water bills before Parliament in connection with which this tendency was strikingly exemplified. A water company which had been supplying the city of Durham from the river Wear since 1847 promoted a bill which was opposed in the House of Lords by the corporation on the ground that the quality of the water was unsatisfactory. Eminent chemists were called in support of the company, and in defence of the Wear. One pronounced the water to be of "unimpeachable quality." Another said it was "an exceedingly good water." A third declared it to be "a very excellent water." A fourth said it was "unexceptionable," and that it was "impossible there could be any sewage contamination injurious to health in the samples examined." A fifth described it as a "very pure water." A sixth called it "an exceedingly good water, without the slightest trace of sewage contamination." On the other hand, it was proved that, above the intake of the company, the river received the drainage of several small towns and villages, containing a population of about 90,000. The House of Lords Committee refused to pass the bill.

The second case is that of the Cheltenham Water Company, who sought to obtain powers to extend their supply from the river Severn. The bill was opposed in the House of Commons by the local authorities.

On behalf of the company, chemical, medical, and engineering evidence was called to establish the wholesomeness of the water, and its fitness for distribution in a town. On the other side there was the usual evidence to show that the river, having received sewage above the proposed point of abstraction, was not suitable as a source for domestic supply. The House of Commons Committee decided that the preamble had not been proved, and advised the company to arrange with the corporation for the transfer of the undertaking.

The mortality returns do not furnish any definite information as to the influence of different classes of water on the death rate of a community. Nor is this surprising when it is remembered how many other influences affecting health are at work.

There are not wanting numerous well authenticated instances in which an improvement in the quality of water has been followed by a marked diminution in the prevalence of disease and in the rates of mortality; but in such instances it has almost invariably happened that other sanitary improvements have been carried out at about the same time, so that it has not been possible to assign to any one specific cause the amelioration that has been effected. If two towns could be found in which the physical and social conditions that affect health were exactly alike, while the water supply in the one was unexceptionable, and in the other admittedly polluted by sewage, a comparison might be made which would perhaps settle the question.

That the use of water which has been in contact with human excreta is productive of disease and death has been only too often proved, both on a large scale, in populous districts, and on a smaller scale, in isolated families. The diseases known to be communicated thus are chiefly typhoid fever, diarrhœa, and cholera. The poison, infectant, germ, parasite, or whatever the material or organism may be, by which the disease is communicated, cannot be detected by the chemist or microscopist, by taste, sight, or smell. No chemist can discover in a sample of water the excreta of a typhoid fever or cholera patient. And yet a healthy person drinking water thus polluted would infallibly be attacked by the particular disease. Chemical analysis may pronounce water safe which is known to produce disease, and at the same time may condemn water which, so far as experience shows, has never caused inconvenience or sickness to the consumer.

The diseases conveyed through water are believed to lurk in certain *bacteria,* or low organisms which have not yet been identified. The existence of such organisms is, however, not a mere matter of speculation, for it is generally admitted that at least splenic fever, pig typhoid, *cholera des poules* or fowl cholera, and relapsing fever, are produced by bodies of this nature, which cannot be distinguished by chemical analysis. Professor Huxley put the question at issue very tersely a few months ago at a discussion on river-waters in the Chemical Society, by asking the president if there was any known method by which, if a drop of Pasteur's solution were placed in a gallon of water, its constituents could be estimated. The president answered that "it was doubtful;" when Professor Huxley went on to show that every cubic inch of that gallon of water would contain from 50,000 to 100,000 *bacteria,* and *one drop* of it would be capable of exciting a putrefactive fermentation in any substance capable of undergoing that fermentation. "For purposes of health," he added, "the human body may be considered as such a substance, and we may conceive of a water containing such organisms, which may be as pure as can be as regards chemical analysis, and yet be, as regards the human body, as deadly as prussic acid."

Chemical analysis can detect the presence of very minute quantities of organic matter in water, but cannot tell much about its character and condition. The opinion pronounced by the Royal Commission on Water Supply in 1869 still applies:—"Where a minute quantity of organic matter escapes destruction, it would seem that chemistry is not yet sufficiently advanced to pronounce authoritatively as to its exact quality and value; and with microscopic living organisms, especially, chemistry is incompetent to deal, and other modes of examination are needed." The Commissioners failed singularly in applying this doctrine, for though they heard twelve chemical and medical witnesses, they did not call a single microscopical witness.

There is undoubtedly a fruitful field of usefulness open for microscopical investigation in connection with water supply.

In pure water, the most powerful microscope cannot

discover the slightest trace of vegetable or animal life. Unpolluted rain and deep well-waters are as free from visible shapes under the microscope as under the unaided eye. Polluted waters, on the other hand, often swarm with low organisms. The absence of organized bodies is no proof that a water is wholesome, nor is their presence conclusive evidence that it is unwholesome. It is the province of the biologist to distinguish and classify all the varieties of visible forms that are present, to trace their development, to discover their *habitat*, and to determine their significance.

After the cholera epidemic of 1854, the committee appointed to make scientific inquiries thought it "a necessary supplement to the chemical inquiry" to obtain a microscopical report from Dr. Hassall on the water supply of the metropolis. It is to be regretted that the same necessity has not been felt in subsequent investigations; for no one can read the able reports of Dr. Hassall to the General Board of Health, in 1850 and 1857, and the Cholera Report of 1854, without recognizing the value and promise of such examinations. The scientific committee above referred to attached "very great importance to the fact, that nearly all the waters consumed in London show a remarkable aptitude to develop low forms of animal and vegetable life." After the epidemic, all the companies drawing their supplies from the Thames removed their intakes to the present stations above Teddington Lock, in accordance with the Act of 1852.

When the new works had come into operation the General Board of Health directed chemical and microscopical examinations to be made to show the effect of the change. Dr. Hassall reported that a "great improvement is undoubtedly manifest in the condition of the present supplies, as shown by the colour and taste of the water, as well as by the diminished number of organic productions contained in them." In his summary of conclusions he stated—

"That the water supplied by the nine Metropolitan Water Companies, under the new Act for the improvement of the water supply of the metropolis, still contains considerable numbers of living vegetable and animal productions belonging to different orders, genera, and species, but especially to the orders or tribes *Annelidæ*, *Entomostraceæ*, *Infusoriæ*, *Confervæ*, *Desmideæ*, *Diatomaceæ*, and *fungi*."

Since 1857 there has been no official microscopical examination of the London supplies.

Some valuable discoveries have lately been made in regard to the relation of micro-organisms to disease, and investigations are now in progress which seem likely to lead to results of still greater importance, bearing upon the origin and spread of infective diseases.

A disease to which wool-sorters are liable has been traced to the *Bacillus anthracis,* derived from the fleeces of animals which have died of anthrax. Water used for washing fleeces and allowed to flow on to pasture-land has been found to produce fatal attacks of splenic fever in sheep and cattle, the *Bacillus anthracis* having been discovered in adundance in their blood.

Among the most recent and suggestive investigations on this subject are those of M. Pasteur, showing the action of a minute form of *bacterium* in producing a disease called *cholera des poules,* which attacked the poultry of Paris during a cholera epidemic; also of Dr. Buchner of Munich, with respect to the relation between the *Bacillus anthracis* and the *bacillus* of hay infusion.*

Apart altogether from chemical and microscopical examinations, the physical conditions of a source of supply afford a good and, in the majority of cases and in competent hands, a safe and sufficient means of determining the suitability of a water for any purpose to which it is

* *Vide* "Untersuchungen über die Actiologie der Wundinfectionskrankheiten," von Dr. Robert Koch. Leipsig, 1878. Also "Ueber die experimentelle Erzeugung des Milzbrandecontagiums aus den Heupitzen," von Hans Buchner. München, 1880. And "Comptes Rendus de l'Académie de Science, Février, Avril, et Mai, 1880."

proposed to be applied. Given the geological features of a watershed or water-bearing rock; the number, distribution, and occupation of the population on the surface; the arrangements for the disposal of refuse; and the character of the water can, by those who are experienced in such matters, be correctly predicated. A broad view of Nature's operations and a just appreciation of Nature's provision for man's wants may not supersede, but should always accompany, the delicate and sometimes finikin processes of the laboratory.

In selecting a source of supply for a town or district, attention should not exclusively be confined to the attainment of a perfect standard of purity. The source that approaches most closely to absolute purity is not necessarily the most suitable in all cases. The conditions of the place to be supplied, the character of the industrial operations carried on, the habits of the people, the proximity and sufficiency of other available sources, must all be taken into consideration. It may often happen that a second-rate water will be better for some places than a first-rate water, by reason of being more plentiful or better adapted to the special wants of the community, or less liable to future contamination. But no difficulty should be allowed to interfere with providing a liberal supply of wholesome water for the domestic uses of every town and district. Freedom from sewage pollution cannot be too strictly insisted upon. At the same time it is as well to remember, in the interests of temperance, that there is probably no supply distributed from public works that can do as much injury to health as the drinks which are generally preferred by the labouring classes.

The only other analytical determination to which I shall refer is the hardness of water. One degree of hardness, according to the standard universally adopted by chemists, is equal to one grain of carbonate of lime, or its equivalent of other hardening salts.

Hardness caused by the presence of carbonates of lime and magnesia is called temporary hardness, because it can be removed by boiling. The carbonate of lime is precipitated, having previously been held in solution by free carbonic acid gas, which the boiling expels. Boiling as generally practised for household purposes has not much effect in diminishing hardness, because the water is either not heated to the boiling-point, or is not boiled for a sufficient length of time to reduce the hardness materially. Hardness due to the presence of sulphates of lime and magnesia is not diminished by boiling, and is therefore termed permanent hardness.

The use of excessively hard water for domestic and manufacturing purposes is objectionable mainly for the waste that takes place in washing and culinary operations, and the injury to boilers and cooking utensils. As to the comparative economy of the two kinds of water, there is no doubt that the advantage is considerably on the side of soft water, especially in places where there are extensive manufacturing concerns; but I do not know of any calculations based upon information systematically collected, showing the consumption of soap, tea, &c., and the wear and tear of linen, in a district supplied with hard water, and in a similar or in the same district supplied with soft water. Calculations based on laboratory experiments and on general statements must always be accepted with reserve.

A chemical commission on the metropolitan supply in 1851 estimated that where soda was avoided a saving of about one-third of the soap used in London for washing linen would be effected by using soft water instead of the ordinary London water, and that the saving in labour would be even more considerable.

The introduction of soft water of 1½ degrees into Glasgow, instead of water of 8 degrees of hardness, was stated by Mr. Bateman, engineer of the Lock Katrine Works, to have resulted in a saving of two shillings per head of the population, to say nothing of diminished wear and tear of clothes in washing. Glasgow manufacturers, who used soap in large quantities, estimated that their consumption was reduced one-half by the change in the water supply. An experienced public analyst has recently made an estimate showing that, allowing one gallon per head per diem for washing and other purposes for which soap is employed, and taking yellow soap at 4*d.* per lb., a saving of £14,600 per annum for every 100,000 inhabitants supplied would be effected in the article of soap alone by introducing a supply of lake water in the place of red sandstone water of about 18 degrees (per 100,000) of hardness. He says, "These figures, far from being exaggerated, are really understated. The consumption of water for washing purposes is much greater than one gallon per head per diem; and one sample of soap bought at a respectable shop was found to contain 69 per cent. of moisture, and was so impure as to be destroyed to the extent of 55lbs. for every degree of hardness in every 10,000 gallons of water."

Every one who has had experience of both hard and soft water for personal ablution, knows how much more pleasant and effective the latter is than the former. With regard to the destruction of boilers and kettles, by incrustations due to deposits from hard water, my experience is that the evil is really serious.

There are many medical authorities who believe that excessively hard water is productive of skin and calculous diseases. There are others who consider hard water beneficial. No evidence in support of hard or soft water is afforded by the mortality returns.

Reference has already been made to the effect of boiling upon the hardness of water. It appears that in order to obtain the full benefit of the reduction which is capable of being produced by boiling, not only must the heat of the water be raised to the boiling-point, but ebullition must continue for about half an hour.

The effect of carbonate of soda in softening water is known to every washerwoman. The bicarbonates and sulphates of lime and magnesia, which are held in solution in the water, are decomposed much more rapidly in hot water than in cold. The carbonate of soda should therefore be applied when the water is hot.

Softening by carbonate of soda is much more economical than softening by boiling, but as the soda salts give an unpleasant taste to the water, this method cannot be employed for dietetic purposes.

The most economical and effective process for removing temporary hardness, that is, hardness due to the presence of carbonate of lime and magnesia, is that known as Dr. Clark's process (from the name of the inventor). By the addition of a certain proportion of lime or lime-water to the hard water, the carbonates are precipitated and a beautifully soft water is produced.

To apply this process on a small scale, I cannot do better than quote the directions given in the sixth Report of the Rivers Pollution Commissioners:—

"To soften 700 gallons of the water supplied by the Chelsea, West Middlesex, Southwark, Grand Junction, Lambeth, New River, or East London Company, slake thoroughly 18 ounces of quicklime (chalk lime is best) in a pailful of water, stir up the milk of lime thus produced, and pour it immediately into a cistern containing at least 50 gallons of the water to be softened, taking care to leave in the pail any heavy sediment that may have settled to the bottom in the few seconds that intervened between the stirring and pouring. Fill the pail again with water, and stir and pour as before. The remainder of the 700 gallons of water must then be added, or allowed to run into the cistern from the supply pipe. If the rush of the water thus added does not thoroughly mix the contents of the cistern, this must be accomplished by stirring with a suitable wooden paddle. The water will now appear very milky, owing to the precipitation of the chalk which it previously contained in solution, together with an equal quantity of chalk, which is formed from the quicklime added. After standing for three hours the water will be sufficiently clear to use for washing, but to render it clear enough for drinking, at least twelve hours' settlement is required."

For the water supplied by the Kent Company about 21 ounces of quicklime would be required to soften 700 gallons of water.

The proportion of lime-water required to soften water from the chalk formations is about 1 to 9, but it varies according to the hardness of the water. The following simple test will enable the operator to know when the proper proportion has been found:—

Take a solution of nitrate of silver in twice distilled water, in the proportion of an ounce per pint. Put two or three drops into a white teacup, and add a little of the mixture of lime and hard water. The tint produced should be a very faint yellow. If the colour is stronger, more of the hard water should be added.

A large proportion of the water used in this country is obtained from the limestone, chalk, and oolite districts, and I have described the application of Dr. Clark's beautiful process somewhat minutely because of its great utility in softening and purifying water from these districts.

The precipitate left after the softening is whitening, which may either be sold or re-burnt for use.

A description of works for softening water on a large scale does not come within the scope of this book, but

4

such works are in successful operation at Caterham, Canterbury, the Colne Valley Waterworks, and other places. The Rivers Pollution Commissioners recommended that the metropolis should be supplied with spring and deep well-water, to be softened with lime before delivery to consumers.

CHAPTER III.

PURIFICATION OF WATER.

Domestic filtration—Filtration on the large scale—Various forms of house filters—Effect of boiling on water.

DOMESTIC filtration is mostly a snare and a delusion. People talk and act as if patent filters were endowed with some magical power for the purification of water. The existence of the impurities which are arrested, if a filter is doing useful work, is forgotten or disregarded. The filth collects on the face and in the body of the filter, and, if the pores are not entirely clogged, an accumulation of putrescent organic matter takes place, which renders the water more impure, though it may be brighter and clearer, after filtration than it was before filtration. The Rivers Pollution Commissioner reported, with respect to one of the best house filters, that "myriads of minute worms were developed in the animal charcoal and passed out with the water when the filters were used for Thames-water, and when the charcoal was not renewed at sufficiently short intervals."

The action of all filters is chiefly mechanical. They are, in fact, strainers, which clear the water of matters in suspension by keeping back solid particles that are too large to pass through the filtering medium. Some filters have also an appreciable chemical action, and remove a portion of the organic matter which is held in solution; but no method of filtration yet devised can remove the ova of animalculæ or deprive water of its power to carry the seeds of disease. A great many varieties and mixtures of material are employed in the construction of filters. The most common are sand, sponge, cotton bags, vegetable and animal charcoal, coke, various preparations of iron, clay, sawdust, flannel, wool, porous stone, wire gauze, gravel. Of these the most effective are animal charcoal, the so-called magnetic carbide of iron, and spongy iron. Animal charcoal is prepared by subjecting bones to a red heat in retorts from which the air is excluded. The material known as "spongy iron" is an iron which has been prepared by the reduction of hematite ore without fusion, and which is consequently in a porous and finely-divided condition. The Rivers Pollution Commissioners, who tested several kinds of filters, found that "fresh animal charcoal removes not only a large proportion of the organic matter present in water, but also a not inconsiderable amount of mineral saline matters." Spongy iron they found to be "a very active agent not only in removing organic matter from water, but also in materially reducing its hardness, and also otherwise altering its character."

Professor Nichols, of Boston, U.S., made an elaborate investigation and report on the filtration of potable water for the Massachusetts State Board of Health. His experiments with an animal charcoal filter gave results similar to those arrived at by the Rivers Pollution Commission. He also obtained favourable results from a silicated carbon filter, described as being composed of the residue of the distillation of a certain variety of bituminous shale compressed into blocks. With a spongy iron filter he experienced a difficulty in obtaining water reasonably free from iron, which he attributed to the Cochituate water (with which he experimented) acting, like other soft waters, violently on iron. As the result of his examinations and researches, Professor Nichols was led to the belief that there is no substance on the whole better than animal charcoal for household filtration.

No filter will continue to give useful results or be fit for use if the filtering material is not regularly renewed or cleansed. The frequency with which this should be done depends on the amount of work the filter has to accomplish. A period of six months is probably as long as a filter in daily use can safely be left without renewal or cleaning.

Water supplied by companies or local authorities ought always to be delivered in a state fit for consumption, without putting consumers to the expense and trouble of household filtration. If filtration is necessary—and it is necessary for the majority of surface waters—it can be most economically and effectively done on the large scale by the waterworks authority. And in all towns and districts where there is a public supply, it would be well if those who now use house filters were to apply their energy and influence to secure the distribution of wholesome water from the public mains to all the inhabitants, rather than trust to a system of domestic filtration which, however excellent it may be in theory, has been found in practice to aggravate the evil it is intended to cure. Any improvement effected in the quality of the general water supply would benefit the poor, who cannot afford to buy filters, as well as the rich, who buy filters and neglect them.

In towns such as Liverpool, Glasgow, and Manchester, where water of unimpeachable quality is delivered to the consumers, house filters are not only unnecessary—they are mischievous. That they have a large sale in such towns shows how ignorant and thoughtless people are on the subject.

For houses that are not supplied from public works

private filtration must often be resorted to, and for the information of those who wish to construct filters for themselves, the directions that follow are given.

There are at least three conditions to be observed in the construction and management of all filters. First, that the filtering material shall not prejudicially affect the water. Secondly, that the size of the filter shall be adapted to the work it has to do, and that water shall not pass through too quickly; and thirdly, that the filtering material shall be periodically cleansed or renewed.

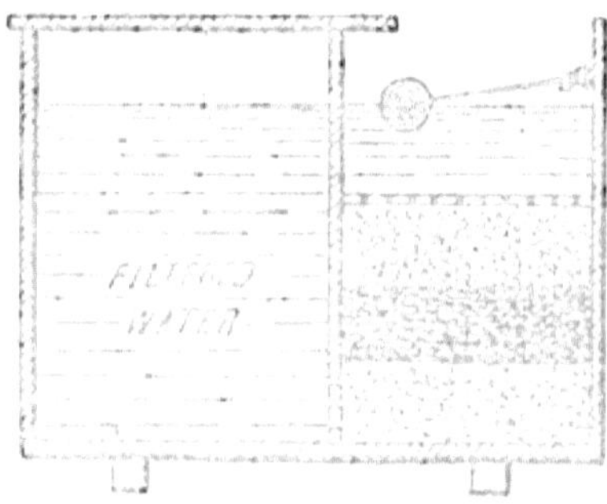

This is a slate or iron cistern and filter combined. The filtering material is, at the bottom, 6 inches of gravel; upon that, 6 inches of animal charcoal; at the top, 6 inches of clean sharp river sand; on the sand a thin brass or tin plate, or slate slab, perforated, is placed to distribute the water. Instead of animal charcoal, the magnetic carbide of iron, or spongy iron, already referred to, may be employed. Failing these, use vegetable charcoal or pounded cinders.

There are endless forms in which the filtering material may be enclosed. Tubs, flower-pots, iron or mug pipes may be utilized for the purpose.

A serviceable filter may be made by letting the water pass through a slab of porous stone, or block of compressed charcoal, sliding in a grove, and dividing the cistern into two compartments.

The following form of filter is adapted for filtering

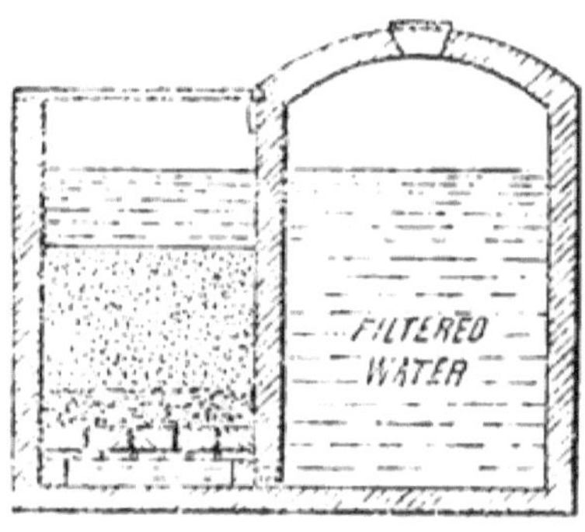

rain-water in an underground cistern, or wherever a large quantity of water has to be filtered. The material proposed is 2 feet 6 inches of sand at the top, then 6 inches of gravel, and below that a course of bricks laid flat on a perforated slab. The tank to be made of brick, lined with cement.

In sand filters, whenever the surface of the sand becomes foul, or the action of the filter slow, scrape off about half an inch, and wash it well before restoring it. Once or twice a year wash all the filtering material. Expose the charcoal to the sunlight for a few hours spread out on a tray, or bake it in an oven. If the charcoal is used in the block form, brush it occasionally.

Effect of Boiling.—Reference has already been made to the effect of boiling in reducing the hardness of water. If there are living organisms present in the water, the value of boiling as a protection to health depends upon its power to destroy vitality. If the temperature of the water is raised to a degree at which such organisms cannot exist, the quality of the water is obviously improved to the extent that inconvenience would have resulted from swallowing them unboiled. As to the temperature at which these things can exist, Doctor Parkes refers to some experiments in which *bacteria* were found moving rapidly at a temperature of 127 degrees C. According to M. Toussaint, the *Bacillus anthracis*, mentioned on page 12, is killed at a temperature of 55 degrees C. (131 F.).

CHAPTER IV.

MODERN WATERWORKS IN ENGLAND.

Origin of water companies--Competition in water supply—History of London supply and its lessons—Duty of local authorities—Services rendered by water companies—Advantages of supply being under control of local bodies.

ABOUT forty years ago, the local administrative bodies of England began to bestir themselves on the subject of water supply. At that time all the waterworks establishments in the country, with a very few exceptions, were in the hands of joint-stock companies. In many instances the works had been first constructed by philanthropic men solely from a desire to provide for a great and pressing public want. In the vast majority of cases the works had been erected simply as speculations by persons who sought a profitable investment for their capital. Thus the supply of water to a town became a trading enterprise, and led to competition that sometimes produced disastrous results. The rates and charges for the commodity supplied were usually fixed by the special Acts of Parliament incorporating the companies, but the limits placed were so high that they were seldom reached. In the earlier Acts there were no restrictions as to the amount of dividend to be paid to the shareholders, nor any provisions to enable consumers to demand a supply. The only protection the public had was their power to set up a rival establishment; and this alternative presented so many difficulties, and involved so much expense in doubtful parliamentary contests, that it was rarely tried. In the few cases where competition did take place, the result was, at first, fierce rivalry, a double expenditure of capital, two or three sets of pipes in one street, two sets of canvassers touting for customers, two sets of officers and workmen, staffs of plumbers to transfer customers from one set of pipes to another, and breaking up of streets; then, when the rival companies had by their competition brought themselves to the brink of ruin, they saw the folly of their ways, and agreed either to amalgamate, or to confine their operations to separate districts, so that in the end the public were worse off, and were burdened with heavier charges, than before the competition began.

As early as 1819 a Committee of the House of Commons came to the conclusion that the principle of competition was not applicable to water companies, but it was not until long after that proper provisions for the protection of the public were inserted in Waterworks Acts.

The history of the supply of London, while it presents many special features, has also some that are common to the supply of other important towns.

The earliest account of the London water supply describes it as being derived partly from the Thames, whence it was carried in butts and buckets through the lanes that ran down to the river-side, partly from the River of Wells, or Wal-brook, which rose to the north of Moorfields, passed through London Wall between Bishopgate and Moorgate, and ran through the City; and partly from a number of springs, the overflow from which ran into the Wal-brook and formed its chief source. As the City extended, the brook had to be covered, the springs were exhausted, and it became necessary to go farther. Water was conveyed to conduits, cisterns, or fountains by earthen or leaden pipes from springs to the north and west of the City. Nearly all these conduits were erected from time to time at the expense of individual citizens, generally Lord Mayors and Sheriffs, anxious to distinguish their term of office by conferring on the inhabitants a boon so highly valued. In 1544 the Corporation of London applied to Parliament for power to convey water from the hills at Hampstead. The works then authorized were not completed until 1590. About that time the corporation obtained power to cut a river for the conveyance of water from any part of Middlesex or Hertfordshire. Ten years were allowed for completing the works, but nothing was done, and the powers lapsed.

The water question appears to have received considerable attention towards the end of the sixteenth century. Among other schemes that attracted much notice and discussion was one to convey water from the springs of Amwell and Chadwell in Hertfordshire. In 1606 the corporation obtained power from Parliament to procure a supply from these sources. The Common Council lacked energy and courage to carry out the undertaking, and it would probably have been abandoned but for the public spirit of Hugh Myddelton, a goldsmith of London, but a native of Henllan in Denbighshire, who undertook to execute the work at his own risk and expense, and to whom the necessary powers were transferred by the Common Council in 1609. According to an edition of Stow's "Survey of London," published in 1633, the obstacles which Myddelton encountered were of a very formidable character. "For if those enemies of all good endeavours, danger, difficulty, impossibility, contempt, scorne, derision, yea, and desperate despight, could have prevailed, by their accursed and malevolent interposition, either before, at the beginning, in the very birth of the proceeding, or in the least stolne advantage of the whole prosecution, this worke, of so great worth, had never been accomplished." Myddelton bravely faced and finally surmounted every difficulty and opposition, and in 1613 had the joy, in which all London shared, of seeing the New River completed. In 1619 James I. granted a charter for the incorporation of the New River Company.

All the works hitherto described have been for supplies by gravitation. In 1582 Peter Morice, a Dutchman, erected a water-wheel at London Bridge, on a lease from the Council, and astonished the good citizens by throwing water over St Magnus's steeple. This was the first pumping machinery used in England. Additional engines were subsequently erected, and the concern developed into the London Bridge Waterworks Company.

On the completion of Myddelton's noble work, the supply of London was thus described:—" What with the spring-water coming from the several springheads through the streets of the city to these cisterns, the New River water from Chadwell and Amwell, and the Thames water, raised by several engines or water-houses, there is not a street in London but one or other of these waters runs through it in pipes conveyed under ground; and from these pipes there is scarce a house whose rent is £15 or £20 a year, but hath the convenience of water brought into it by small leaden pipes laid into the great ones. And for the smaller tenements, such as are in courts and alleys, there is generally a cock or pump common to the inhabitants; so that I may boldly say, there is never a city in the world so well served with water."*

The success of the New River led to several new schemes being brought forward, and to the establishment of several new companies. The next important event in the history of the London supply was the erection of a steam engine by Savory, at York Buildings, under "the governor and company of undertakers for raising Thames-water in York Buildings," incorporated in 1691. About half a century before, water had been raised from the Thames at Vauxhall by an engine invented by the Marquis of Worcester. In 1723 the Chelsea Waterworks Company was formed. They pumped from the Thames at Chelsea Reach, and had power to use and make ponds or reservoirs in St. James's Park and Hyde Park. In 1785 the Lambeth Company was formed. They pumped from the Thames near Waterloo Bridge. In 1806 the West Middlesex Company was incorporated, to supply the northern district of London. They pumped from the Thames at a point about 9½ miles above London Bridge. In 1798 the Grand Junction Canal Company obtained power to construct waterworks to supply Paddington and the adjacent parishes. In 1811 the powers thus obtained were transferred to a

* Stow. Edition 1633.

number of persons who were incorporated as the Grand Junction Water Company. Their supply was at first obtained from the Grand Junction Canal, fed from the rivers Colne and Brent; subsequently a supply from the Regent's Canal was added; but in 1820 both these sources were abandoned, and the whole supply was derived from the Thames at Chelsea. In 1807 the East London Company was incorporated to supply water from the river Lea at Old Ford to districts in the east of London, previously supplied by the Shadwell and West Ham Waterworks. In 1822 the Borough Works, at St. Mary Ovary's, and the London Bridge Works were united to form theSouthwark Waterworks. In 1805 the South London Company was established, and erected engines on the Thames at Cumberland Gardens, near Vauxhall Bridge. The Southwark Company and the Vauxhall Company were amalgamated in 1845. In 1809 the Kent Company was formed, to supply a district south-east of the Thames, from the river Ravensbourne (abandoned in 1862 in favour of deep wells in the chalk).

Thus five new companies were established about the beginning of the present century. Several of these were competing companies, and the rivalry between them was keen and bitter. Each company tried to undersell the others. Canvassers went about trying to persuade people to transfer their support from one company to another.

The inducements commonly offered were; a reduction in price, greater liberality and regularity in supply, and improvements in pressure. The competition continued with vigour, and not a little acrimony, for many years, and while it lasted the public derived considerable benefit from it. After a great waste of capital and loss of dividends, the competing companies saw impending ruin as the end of the struggle, and they came to an agreement that each company should confine its distribution to certain specified limits, and not interfere with its neighbour. The New River and East London Companies were the first to suspend hostilities, in 1815. The other companies to the north of the Thames came to an understanding in 1817, and the companies south of the river in 1812.

The first result of the treaty between the companies was an increase in the charges for water. A great outcry was raised against this action, and an association was formed, called the "Anti-Water-Monopoly Association," to resist the increase. In consequence of this agitation a Select Committee was appointed by the House of Commons in 1821 "*to inquire into the past and present state of the supply of water to the metropolis, and the laws relating thereto.*" The Committee took evidence as to the effects of the mutual compacts agreed upon by the companies on the Middlesex side, and they reported that legislative regulation was required. In the same session a bill was introduced to appoint referees to decide any questions of dispute that might arise between the public and the companies. The bill failed to pass.

At this time the quality of the Thames water began to receive attention, and in 1827 a Commission was issued appointing Mr. Telford, the eminent engineer, Dr. Brande, F.R.S., and Dr. Roget to inquire into the "quality and the salubrity of the water." In their Report the Commissioners made the following remarks in reference to the inquiry by the Select Committee of 1821: "The disposing of such a necessary of life ought not to be left altogether to the unlimited discretion of companies possessing an exclusive monopoly of that commodity, and the interests of the public require that while they continue to enjoy that monopoly their proceedings should be subject to some effective superintendence and control." With regard to quality, the Commissioners reported that it was susceptible of, and required, improvement; "that it ought to be derived from other sources than those now resorted to, and guarded by such restrictions as shall at all times insure its cleanliness and purity."

In 1831 the Government instructed Mr. Telford to report "in what manner the metropolis can be supplied with pure water." Mr. Telford's report was issued in 1834. He recommended a scheme for supplying the districts to the north of the Thames from the river Verulam, and the districts to the south from the river Wandle. Numerous projects for an improved supply were brought forward about this time, and attempts were made to form new companies, but none of these proposals went beyond the speculative stage. The water companies, having ceased to compete with each other, gave more attention to the improvement of their works. The Chelsea company constructed a filter bed in 1829, and this was the first attempt to improve the Thames-water by artificial filtration.

I now come to the time when the Health of Towns Commissioners made their inquiry (1844). In regard to London, they found "the defects in the system of distribution and charging frequently more striking" than in other towns. "Large numbers of the houses of the poorer classes receive no supply." Out of 900,000 persons in the New River Company's district, 300,000 were "unsupplied," and in the district of the Southwark Company 30,000 had no supply. Large numbers of the inhabitants, especially of the poor, were dependent on shallow wells, and "the practice, almost universal, of retaining all refuse in cesspools beneath houses, has, in many parts of the metropolis, so entirely saturated the soil as to render the water unfit for use."

The Commissioners thought that the principles of

legislation which they had recommended for other large towns were equally applicable to the metropolis.

In 1848-9 London was visited by an epidemic of cholera. Some medical men attributed its spread to the agency of water, and this led to general uneasiness on the subject. In 1850 the General Board of Health issued an elaborate Report, in which the question was very fully examined from an engineering, medical, and chemical point of view. The result was that they condemned unreservedly the use of Thames water taken within reach of London sewage, and condemned the Thames beyond the influence of London sewage, and likewise the Lea and New River, on account of their hardness. During the Board of Health inquiry no fewer than twenty-five new projects for improving the supply to the metropolis were submitted. In 1851 another Commission was appointed, to report specially upon the chemical aspect of the question. The Commissioners were Professor Graham, F.R.S., Dr. W. Allen Miller, M.D., F.R.S., and Dr. Hofmann, F.R.S. They prepared a very able Report, in which, after praising the Thames, with ordinary sand filtration, for its "peculiar and agreeable brightness," they went on to say, "As the main drain of a large and populous district, the Thames becomes at all seasons polluted by the sewage of several considerable towns, and by the surface drainage of manured and ploughed land. Even above Kingston a population of three-quarters of a million is found upon the banks of this river and its tributaries. The diverting the sewage of the various towns entirely from the Thames would be attended with so much difficulty that the project need not be taken into account. . . And it appears to be only a question of time when the sense of this violation of the river purity will decide the public mind to the entire abandonment of the Thames as a source of supply; unless, indeed, artificial means of purification be devised in the meantime and applied." To the Lea they referred in much the same language. In consequence of these Reports the Government introduced a bill in 1851 to affect an amalgamation of the companies, and to enable the Secretary of State to prescribe the sources from which water should be obtained. This bill was strenuously opposed by the companies, and had to be withdrawn. In 1852 an Act was passed "to make provision for securing the supply to the metropolis of pure and wholesome water, and otherwise to make further and better provision in relation to the water supply of the metropolis." These objects it sought to attain by enacting: Firstly. No company to take water from any part of the Thames below Teddington Lock, or from any of its tributaries below the point where the tide flows. Secondly. All water, except the deep well-water, to be effectually filtered before distribution. Thirdly. All reservoirs for filtered water within five miles of St. Paul's to be covered. Fourthly. No company to resort to any new source of supply until an inspector of the Board of Trade shall have reported if the same is capable of supplying good and wholesome water. Fifthly. On any complaint from twenty householders as to quantity and quality, Board of Trade may appoint a person to inquire and report. Sixthly. If complaint well founded, Board of Trade to require company to remove grounds of complaint. Seventhly. After expiration of five years, every company to give a constant supply, at high pressure, of pure and wholesome water, "provided that no company shall be bound to provide a constant supply of water to any district main until four-fifths of the owners or occupiers of the houses on such main shall have required such company to provide such supply, nor even upon such requisition in case it can be shown by any company objecting that more than one-fifth of the houses are not supplied with pipes, cocks, cisterns, machinery, and arrangements of all kinds for the reception and distribution of water according to the prescribed regulations."

The first effect of the Act of 1852 was the removal of the intakes of the Grand Junction, West Middlesex, and Southwark and Vauxhall Companies to Hampton, and of the Lambeth and Chelsea Companies to Thames Ditton.

In 1856 the Board of Health directed Professor Hofmann and Mr. Lindsey Blyth to report upon the quality of the water from the new intakes. They reported a very considerable diminution in the amount of organic matter as compared with 1851, but expressed no opinion as to the salubrity of the water examined. In the same year the superintending inspectors of the Board were required to report on the alterations effected in the works under the Act, 1852. They reported that the requirements of the Act, in regard to sources of supply, filtration, and covering of reservoirs, had "in all essential respects been fully and satisfactorily complied with." The inspectors, having regard to their experiences under the Public Health Act, were led to consider whether it might not be desirable to give early attention to the number of towns situated above the new sources of supply, whose drainage might, sooner or later, "to some extent render nugatory the vast outlay which the companies have been called upon to make." The five years allowed for the introduction of constant service had not then expired.

In 1857 the Board of Health obtained a Report on a "Microscopical Examination of the Metropolitan Water Supply," by Dr. Hill Hassall. The Report stated that "the metropolis is still supplied with water containing considerable numbers of living vegetable and animal productions, and which are not present in the purer waters," as from deep wells in the chalk.

In 1865 the first Rivers Pollution Commission was appointed. A Report on the Thames and Lea was issued in 1866. Concerning the Thames the Commissioners remarked, "The result seems to be that as a water supply the Thames, polluted by the sewage of the inhabitants of the river basin, is open, in kind if not in degree, to the same objections as well-water infiltrated by liquid from an adjoining cesspool. Well-water so tainted may appear to sight, taste, and smell to be harmless, and has been known to have been drunk for a length of time without apparent mischief, but beyond all doubt that same water is liable under perculiar conditions to become poisonous." As to the Lea, they reported, "we have found on inquiry that pollution of the waters by sewage is general."

In 1867 a Committee of the House of Commons on East London Water Bills reported as follows: "We are satisfied that both the quantity and quality of the water supplied from the Thames were so far satisfactory that there is no ground for disturbing the arrangements made under the Act of 1852, and that any attempt to do so would only end in entailing a waste of capital and an unnecessary charge upon the owners and occupiers of property in the metropolis."

In 1867 a Royal Commission was appointed to inquire generally into the subject of water supply, and to report which sources "are best suited for the supply of the metropolis and its suburbs." The Commissioners reported (1869) "That there is no evidence to lead us to believe that the waters now supplied by the companies is not generally good and wholesome," and "that when efficient measures are adopted for excluding the sewage and other pollutions from the Thames and the Lea, and their tributaries, and for insuring perfect filtration, water taken from the present sources will be perfectly wholesome and of suitable quality for the supply of the metropolis. That the constant service system ought to be promptly introduced to the furthest extent possible in the metropolis. That the future control of the water supply should be entrusted to a responsible public body, with powers conferred upon them for the purchase and extension of existing works."

In 1871 the Government introduced a bill for the compulsory purchase of the water companies. This bill met with the usual fate of measures to deal effectually with the London water question,—it had to be withdrawn. A bill (No. 2) was introduced in the same session, the object of which was thus described by the Select Committee to whom it was referred:—To secure (1) distribution of water on a system of constant supply.

(2) Good quality of the water. (3) An audit of the companies' accounts.

The Metropolis Water Act, 1871, sought to accomplish what the Act of 1852 had failed to do in regard to constant service by empowering the local authorities of the metropolis (subject to an appeal to the Board of Trade) to make the demand for constant service instead of the inhabitants. In other respects the conditions remained much the same.

In 1874 the Rivers Pollution Commissioners issued their Report on the "domestic water supply of Great Britain," in which they discussed at great length the metropolitan supply. They made the following recommendations: "That the *Thames* should as early as possible be abandoned as a source of water for domestic use, and that the sanction of your Majesty's Government be in future withheld from all schemes involving the expenditure of more capital for the supply of *Thames* water to London." "That the *Lea* should also be abandoned as a source of potable water," but "this measure is less urgent than the relinquishment of *Thames* water."

The next Parliamentary inquiry relating to the London water question was that of a Special Committee on the Metropolitan Fire Brigade, which sat in 1876-7. They recommended "That the water systems now belonging to the various companies should be consolidated in the hands of a public authority, which, in dealing with the questions of constant supply, pressure, and pipeage, should be bound to have regard not only to the convenience of customers, but also to the requirements for the extinction of fire."

In 1878 the Metropolitan Board of Works promoted two water bills; the first, For the acquisition of the water companies' undertakings; the second, For providing a new supply for potable and fire-extinquishing purposes. Neither of these measures got beyond the stage of a first reading. The introduction and abandonment of the bill introduced this year by Government for the purchase and transfer to a public authority of the water undertakings is within the memory of all. The Select Committee subsequently appointed to consider the proposed terms of purchase arrived at conclusions similar in effect to those of the Royal Commission of 1814.

The history of the London water supply corresponds up to a certain point with that of many towns, and furnishes lessons that are valuable to all. First, there was a period when the inhabitants were dependent on shallow wells, on water drawn by hand from the Thames, and on the munificence and public spirit of individual citizens. Then came a period when the growth of the city demanded more extensive provision, and the adoption of more comprehensive measures. The city council were not sufficiently advanced in their conception of the functions of a local administrative body, and had not sufficient enterprise to carry out so large an undertaking. Hugh Myddelton came forward to perform the task from which the council shrank. Thenceforth a responsibility which ought to have devolved on the citizens as a corporate body was left to speculators. An era of competition began. Rivalry was carried to the verge of ruin. Then peace was declared, and a monopoly established by an agreement in which the wishes and interests of those chiefly co.icerned were never consulted. Then followed a period of barren public agitation and of bubble projects. After this an epidemic of cholera aroused attention. Medical men pronounced water to be a carrier of the disease, and people who learnt that their supply was drawn from a river at a point within a few yards of a sewer-outlet were naturally alarmed. Government was compelled to interfere, and after another futile attempt to cope with the difficulty by putting an end to a bad system, they gave a new lease to monopoly. Then we come to a period in which sanitary science has been claiming more and more attention, and new and higher standards of purity are proclaimed ; when the evils of a defective system have become intolerable, and general uneasiness prevails regarding the wholesomeness of water derived from a river which has received the filth of a large population.

That it is one of the first duties of a local administrative body to provide an adequate supply of water for the inhabitants of their district, is a proposition which is now almost universally established; and Parliament has since 1846 acted on the principle that it is desirable to have public waterworks under the control of local authorities, to be managed by them for the benefit of the ratepayers. Local authorities have obvious advantages over joint-stock companies, in connection with the breaking up of streets for pipe-laying, the collection of rates, the enforcement of regulations, in matters of policy, and in being able to borrow money at low rates of interest; but the services of water companies ought not to be overlooked. We should not forget how much the country is indebted to private enterprise, often stimulated quite as much by a desire to render useful service as by an expectation of profit, for many valuable works that are scattered over the country. Nor should we forget the numerous improvements that have been effected, by and through water companies, in machinery and appliances of various kinds, and in methods of distribution; and the examples they have furnished of economic working and judicious management. Especially is it worthy of recollection that it is to a water company, acting under the advice of Mr. Hawksley their engineer, that we owe the first introduction of constant service at high pressure. There are many advantages which a company possesses over an elective body,—in unity and continuity of purpose and action, in conducting negotiations, in personal supervision, in freedom from influences, restraints, and political ties, which often embarrass and determine the conduct of public men. Some of the best waterworks in the country are in the hands of private companies, who are supplying their districts with wholesome water on reasonable and even liberal terms.

Notwithstanding all this, the balance of advantages is decidedly in favour of waterworks being in the hands of local administrative bodies, directly responsible to, and representative of, the consumers, and charged to see, not that a maximum dividend is obtained, but that there shall be no manner of doubt or mistake as to the purity of the water, and its sufficiency for all the domestic and trade requirements of the poorer as well as of the wealthier classes, and for all public and sanitary purposes.

CHAPTER V.

DISTRIBUTION OF WATER.

Constant and intermittent systems—Evils of intermittent system—Consumption of water under both systems, and in various classes of property—Liverpool experience—Proposed supply of houses through meters—Objections to unnecessary restrictions—Correct principles of distribution.

A CONSTANT supply of water is defined by the Waterworks Clauses Act as a supply "constantly laid on at such a pressure as will make the water reach the top storey of the highest houses within the limits" of the special Act.

In districts where water is supplied according to this system the consumer can obtain water from a tap in direct communication with a street main at any time, except when the main itself is shut off for repairs, alterations, or extensions.

Where the service is intermittent the water is generally turned into the distributing mains every day for a length of time, varying from one to eighteen hours.

The evils attending the intermittent system are many and serious.

(a.) While the supply is shut off, water can only be obtained from cisterns, butts, and other vessels, in which it has been stored while the mains are charged. The erection and maintenance of these receptacles involves considerable expenditure of money, and the water stored is often polluted by floating particles from the atmosphere, and by the absorption of noxious gases. With a constant supply, storage cisterns are only necessary for boiler supplies and trade purposes.

(b.) The impurity of cistern-water leads to the frequent use of domestic filters, which are seldom properly attended to, and therefore only serve to aggravate the evil they are intended to remedy.

(c.) The poorer classes, who are not adequately provided with cisterns, are compelled to use kettles, pans, jugs, and other household utensils, which are generally exposed to the vitiated air of overcrowded rooms, and consequently cannot preserve water in a state fit for consumption.

(d.) When the water is turned off, the mains are more or less emptied, and a vacuum or partial vacuum is formed in them. The consequence is that foul gases

and even liquid sewage may get into the mains, through defective pipes and open taps.

(e.) Dwelling-houses, and especially the houses of the poor, become damp and unhealthy, owing to leakages soaking into the foundations from faulty pipes. Such leakages would not be allowed to continue under a well-managed constant supply system.

(f.) Except during the hours of distribution, and unless special fire mains are provided, no water can be obtained to extinguish fires until a turncock or fireman has opened a valve communicating with a charged main or reservoir. This operation causes delay at the most critical time in the progress of a fire.

Further, when the service-valve has been opened, the waste of water through open taps and imperfect pipes in the neighbourhood is so great that the pressure is too feeble to be available for putting out a fire without the aid of a pump. The value of an adequate and constant supply of water as a means of extinguishing fires has been shown by the great reduction of insurance rates in Manchester, Liverpool, and other places, on the substitution of a constant for an intermittent service.

(g.) While the useful consumption of water is restricted, the amount of waste is greater under an intermittent system than under a properly regulated constant system.

Where steam power is employed for pumping, as in London, the cost of procuring the water increases in proportion to the waste. The average consumption in London during the year 1879 was 32½ gallons per head per day. Fully one-half of this water was wasted. Not that the whole of the waste could be stopped by the introduction of a well-managed constant supply, but there would not be much difficulty in effecting a reduction to the extent of one-third of the present consumption.

(h.) Under the intermittent system a much larger staff of turncocks has to be employed than under the constant system.

(i.) Where only a short or irregular intermittent service is given, the pressure from the mains cannot be used for motive power.

The statement that more water is consumed when a supply is turned on for only a few hours daily than when it is constantly on may require further explanation.

If we suppose the case of water flowing from a tap under an invariable head, the quantity used would obviously be in proportion to the time during which the tap was open. There are places in which the pipes and fittings are so defective that the conditions are almost analogous to the open tap case. But where a constant supply is under proper control, and where judicious regulations for the prevention of waste are enforced, the consumption of water is less—and often very considerably less—than in places where the supply is intermittent. Perhaps this im-

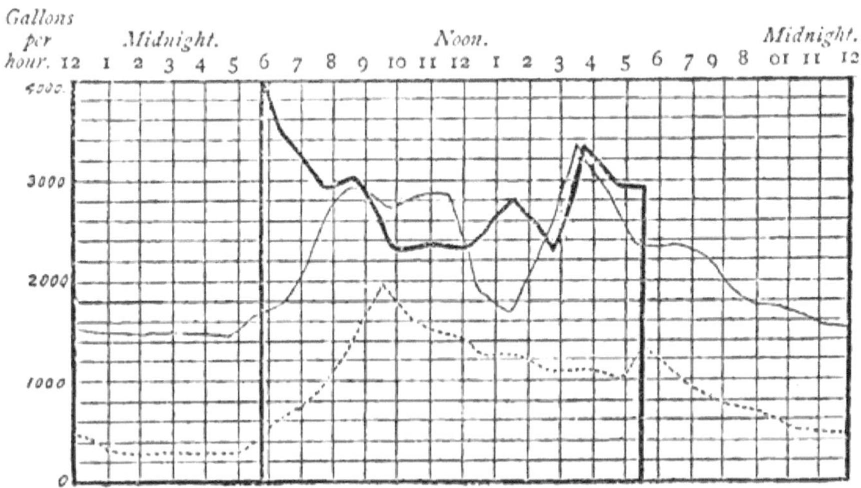

portant point will be made more clear by an illustration from a district comprising several streets in the middle of a town in which the supply passed through the three stages of intermittent, wasteful constant, and economical constant service. The annexed diagram shows the number of gallons flowing into the district each hour under the several conditions named. The thick black line represents the intermittent supply; the thin black represents the wasteful constant supply; the dotted line represents the economic constant supply. It will be observed that during the intermittent service period the water was turned on at 5.50 a.m., and shut off at 5.30 p.m.

The total consumption in this particular case was:—

Intermittent supply ...	18·8	gallons per head per day.
Wasteful constant supply..	29·3	do.
Economic constant supply	12·4	do.

During the intermittent service to which the diagram relates, the pipes and fittings were periodically examined by inspectors going from house to house, and serving notices to repair all leakages which they discovered. It is, therefore, not a case where the waste was due to exceptional neglect; nor is it to be regarded as being, in the relation between the columns, or indeed in any sense, a representative case. The consumption per hour, in any district or town, increases as the total number of hours supply per day decreases; and the inconvenience and danger to which consumers are exposed increase and decrease in precisely the same order and proportion.

An interesting and instructive example, on a large scale, of the relation between the length of time during which water is turned on, and the total daily consumption, is afforded in the case of Liverpool. From the year 1858 Liverpool had a constant service until the summer of 1865, when, in consequence of an unusual drought, it became necessary to restrict the supply. At first the length of the daily service was reduced to seven hours. The duration of the supply was therefore diminished to the extent of 70 per cent.; but the diminution in the total daily consumption, resulting from the change, was only 5 per cent. And this notwithstanding that considerable alarm prevailed in the city from the fear of a water famine, and notwithstanding appeals made to the inhabitants to exercise every possible economy. The supply was afterwards restricted to four hours per day, and this further restriction produced a diminution of 28 per cent. Throughout the year 1866 the intermittent service was continued, and the average consumption per head was 21 per cent. less than obtained in 1864 with a constant service.

Year.	No. of hours Water turned on daily.	Average No. of galls. consumed per head per day.	Average No. of galls. consumed pr. hour of supply
1864	24	30·42	1·26
1865	Period of	transition.	
1866	10	23·85	2·38
1867	14¼	28·24	1·98
1868	11¼	25·90	2·30
1869	11	27·51	2·50
1870	10	27·94	2·79
1871	9½	29·35	3·09

The preceding table gives the rate of supply from year to year.

The gradual increase of waste under the intermittent system is very clearly shown. Between 1866 and 1871 the total consumption per head per diem increased 23 per cent., and the rate per hour increased 30 per cent. Throughout this period a systematic house-to-house inspection was maintained. In 1873 special operations for the prevention of waste and restoration of constant service were commenced, and by the end of 1875 a constant supply had been restored to the whole of the city. The rate of consumption under the constant system is now 22½ gallons per head per day. This change has been accomplished, and this reduction effected, without any sweeping interdict of old and faulty fittings.

The average volume of water required, per head of the population, to give an adequate supply to a town, depends upon local conditions, in regard to which no general rule can be laid down. The rate varies from 10 to 40 gallons per head per day, of which a large proportion is waste, which no regulations or supervision can wholly suppress.

The quantity used in different classes of property in the same town also varies considerably. In the annexed table I give a few examples of the consumption for domestic and miscellaneous purposes in various classes of property in a large town, where water-closets are in general use, and the supply is constant.

	Population of District.	Total consumption per head per day. galls.
District containing lowest class of old cottage property, chiefly supplied by outside self-closing standpipes	2,130	7·0
District containing a low class of old houses, occupied by labouring population, mostly supplied by a separate tap in each dwelling	5,800	11¼
District of old houses, small and middle-class, with numerous shops	6,100	11¾
Another ditto	5,300	14
District containing new cottage property, chiefly occupied by clerks, artizans, and small shopkeepers	7,280	12
Ditto	2,300	14½
Ditto	3,900	10¼
District containing average middle-class and small cottage properties	5,400	17¼
District containing best class of town houses (out of London)	2,100	18¼
Ditto	2,000	21
Suburban district containing large houses with gardens	3,450	30
Ditto	990	50¾

The consumption in the districts comprised in this table was obtained by measuring the water flowing into each district through the distributing main; the figures therefore include waste. Trade supplies through meters are not included.

In court houses occupied by the labouring classes the consumption is about 4½ gallons per head per day when the supply is drawn from an outside standpipe, and from 5 to 15 gallons per head per day when there is a separate tap in every house. These figures include water for water-closets common to all the inhabitants of the courts. In large houses, where baths are freely used, the consumption often reaches 70 gallons per head per day. These rates per head are taken from actual measurement of the water flowing through the house-service pipes.

The quantity used for manufacturing and other trade purposes varies so considerably, and there is so much diversity of practice with regard to the employment of meters and charging by assessment, that no useful comparison can be made between the rate in any two towns or districts without some knowledge of the local circumstances.

A total rate of 25 gallons per head per day is a liberal allowance for all the domestic, trade, sanitary, and public purposes (including waste) of a town in which there are considerable, but not exceptional, manufacturing industries, and in which water-closets are not in general use.

In water-closet towns, 30 gallons per head is a liberal quantity to allow for all purposes.

The following are some examples of the present rates per head per day in various towns, for all purposes:—

EXAMPLES

OF THE PRESENT CONSUMPTION OF WATER IN VARIOUS TOWNS IN THE UNITED KINGDOM AND IN AMERICA.

I.—*IN GREAT BRITAIN AND IRELAND.*

INTERMITTENT SYSTEM.

Name of Town or District.	Estimated Population supplied in 1880.	Average consumption in gallons per head per day, for all purposes.		
		Domestic.	Trade, &c.	Total.
Barnet	8,720*	—	—	25
Bath	52,535*	—	—	25
Birkenhead	59,000	—	—	29½
Brighton (partly constant)	130,000	—	—	29
Ealing	9,500*	—	—	20
London	4,289,544	—	—	32½
Oxford	34,182*	—	—	30
Tranmere	23,100	17	2	19

CONSTANT SYSTEM.

Name of Town or District.	Estimated Population supplied in 1880.	Domestic.	Trade, &c.	Total.
Aberdeen	100,000	—	—	45
Ashton-under-Lyne	50,000	—	—	20
Barnard Castle	4,200*	—	—	30
Birmingham	500,000	—	—	20 to 25
Blackburn	100,000	—	—	25
Bolton	200,000	—	—	25
Bradford	183,000	24	16	40
Brecon	5,845*	—	—	15
Bury	100,210	—	—	20

CONSTANT SYSTEM *continued.*

NAME OF TOWN OR DISTRICT.	Estimated Population supplied in 1880.	Average consumption in gallons per head per day, for all purposes.		
		Domestic.	Trade, &c.	Total.
Carlisle	35,000	—	—	20½
Chorley	18,000	—	—	12½
Cockermouth	5,115*	—	—	26
Dawdon	9,031*	—	—	12
Dublin	246,300	—	—	38
Dundee	171,000	—	—	35
Ebbw Vale	12,000*	—	—	15
Edinburgh	304,000	—	—	40
Glasgow	760,000	42¼	7¾	50
Halifax	140,000	—	—	26
Hull	150,000	—	—	32⅔
Kirkcaldy and Dysart	27,000	25	9½	34½
Leeds	312,000	—	—	23¼
Liverpool	703,000	15	7½	22½
Manchester	900,000	13	7	20
Norwich	70,000	—	—	16
Nottingham	186,970	11½	7¼	18¾
Perth	27,000	—	—	28
Plymouth	80,000	—	—	49
Preston	100,000	—	—	20
Ripon	6,80 *	—	—	25
Sheffield	300,000	—	—	18
Southport	21,204	—	—	25
St. Helen's	60,000	14¾	16¼	31
Wolverhampton	96,000	18	3	21
Worcester	33,000	—	—	26⅓
York	60,000	—	—	23

The examples marked thus * are taken from the "Urban Water Supply" Return, Session 1879, and the population given is that of 1871.

II.—*IN AMERICA.*

City.	Year.	Population.	Total consumption in imp. galls. per head per day.
Boston	1877	390,000	62⅓
Brooklyn	1877	485,000	52½
Cambridge	1877	48,000	46
Cincinnati	1877	280,000	47⅓
Chicago	1877	440,000	99
Ditto	1878	—	101½
Cleveland	1877	135,000	46½
Milwaukie	1877	130,000	44
Montreal	1877	130,000	57½
New York	1880	1,208,000	65
Philadelphia	1877	817,000	48½
Providence	1877	100,000	21
Rochester	1877	82,000	29
St. Louis	1877	400,000	46½
Toronto	1877	75,000	64
Washington	1878	143,518	138

A daily average of the consumption during a year does not show the full extent of the demands which waterworks must be adapted to meet. The consumption varies greatly from day to day and from week to week, according to the domestic and industrial habits of the consumers, and according to the state of the weather. In hot summer weather the consumption is about 20 per cent. more than the average of the year. This increase arises from street-sprinkling, garden-watering, use of baths, fountains, and a more lavish employment of water in every direction. A severe frost taxes the resources of waterworks more heavily than a high temperature. The increased consumption caused by frost is often from 30 to 40 per cent. above the average. Taps left open to prevent freezing in the pipes, and the bursting of pipes by the action of frost, account for this.

It has frequently been proposed, and in the United States of America an attempt has lately been made, to control and check waste by applying a meter to every house, charging by measure for the water used for domestic purposes. There are numerous objections to this method of supply, but perhaps the chief objection is that its adoption would certainly tend to restrict unduly the legitimate and necessary use of water. It need scarcely be said that social improvement depends to a great extent upon, or must, at least, be concurrent with, a more liberal use of water, and the promotion of habits of cleanliness. Any arrangement which enhances the price, or in any way checks or discourages the free application of water to useful purposes, is an obstacle to social progress.

The cost of the house-meter system is also a serious objection to its introduction. A good water meter suitable for a domestic supply cannot be bought for less than from £3 to £5. If we take into account the amount of capital which would have to be expended in the purchase of meters, the cost of fixing them and making alterations in existing arrangements, the reproduction fund required to replace the meters when worn out, and the cost of inspection, repairs, and bookkeeping, we shall find that the expenditure per house would be nearly equal to an average annual water rate.

Then there is the further consideration that this expense, and all the restriction, annoyance, and interference involved in the system, are totally unnecessary for the attainment of the object in view. By the aid of modern appliances waste can be very easily detected, and can, without extravagant expenditure or vexatious surveillance, be kept within reasonable limits.

It is highly important that this should be clearly understood for the information and guidance of those who are suffering the inconvenience of a restricted supply, that they may exert their influence in the direction of improvement; and also to show how necessary it is that every consumer should co-operate with the waterworks authorities in preventing undue waste, and in carrying out such regulations as experience has proved to be effective in maintaining an economic constant supply.

CHAPTER VI.

WATER RENTS, RATES, AND CHARGES.

Domestic water rents—How levied—Extra charges—London rates—Charges for sanitary supplies.

THE rates charged for domestic supplies are generally lower in places supplied by local authorities than in places supplied by private companies. A table is appended containing numerous examples of the amounts actually charged in different parts of the country, and under various kinds of management. There are three points in regard to which caution is needed in comparing rates as specified in Acts of Parliament, or as set forth in scales of charges.

The first is as to whether the domestic water rate, or rent, is levied upon the net annual value, or upon the gross annual value; upon the poor rate assessment, upon the rack-rent, or upon the full rent. The difference may be very considerable, as the deductions made from the rent in arriving at the rateable value are much greater in some places than in others. To illustrate this, I will assume that in three towns the domestic water rate is 1*s*. in the £1: the actual water rate charged upon a house let at a rental of £100 per annum may differ in the three towns thus:—

	Amount of Water Rate per annum.		
In *a*, rent £100, domestic water rate charged on rent.	£5	0	0
In *b*, rent £100, domestic water rate charged on net annual value, deducting 10 per cent. off rent	4	10	0
In *c*, rent £100, domestic water rate charged on net annual value, deducting 30 per cent. off rent	3	10	0

The second point to be observed is with respect to extra charges. There are waterworks authorities whose domestic water rates appear to be very low, but who more than compensate for the moderation of the rate by imposing special charges for water-closets, baths, gardens carriages, and other miscellaneous purposes. A definition of what is included in a "domestic supply" is to be found in the special Act of every waterworks undertaking, and for any use of water that is not strictly "domestic," as defined by the Act, the undertakers may make any demand they choose. In the case of London, there is a further source of misconception, in the shape of an extra charge varying from 10 to 25 per cent. upon the statutory rates for "high service;" meaning a delivery of water at elevations varying from 10 to 20 feet above the adjoining pavement. To show clearly the effect of these extra charges, I give a few examples of rates at present levied for houses let at £10 and £150 per annum, the water being delivered into cisterns at the tops of the houses.

		House at annual rent of £40, with one water-closet and one bath.			House at annual rent of £150, with two water-closets, two baths, carriage and two horses.		
		£	s.	d.	£	s.	d.
Liverpool...........	Domestic rent, 6d., Public rate, 6d. in £ on rateable value	1	16	0	6	15	0
Suburbs of ditto............	Domestic rent, 9d.	1	5	6	4	15	3
London—(East London Co.'s district)	5 per cent on annual value, and 25 per cent. extra for high pressure......	3	0	0	12	7	6
Ditto (Chelsea Company)	4 per cent. on annual value	2	4	0	7	12	0
Glasgow	Domestic rent, 8d. in £ on full annual value	1	6	8	5	0	0

Both in Liverpool and in the East London Company's district the amount of water rate payable in respect of a house is 5 per cent. on the annual value. But one is the net and the other the gross annual value, and by the addition of special charges in the one case, which are not made in the other, the tenant of a London house at a rental of £40 pays 33 per cent. more than the tenant of a similar house in Liverpool, omitting from the calculation the extra charge for high service; including the high service, the London occupier pays 66 per cent. more than the Liverpool occupier. Dealing in the same way with a house at a rental of £150, with the additions named in the table, the excess paid in London without high service is 50 per cent., and with high service 83 per cent. Beyond this the charge in Liverpool includes payment for water for sanitary and other public purposes, which the London charge does not include. In Glasgow the difference, as against London, is still greater.

The third point, to which attention is to be directed in comparing the water rates of different towns, is the extent to which the yield of those rates covers the expenditure incurred in connection with the waterworks. Supplies for public sanitary purposes are often paid for out of general district rates, or by special water rates. Deficiencies in the water account are also sometimes charged to a district rate. In some instances, the public water rate and the domestic water rent are treated as one account, and not separately applied. Where the waterworks are in the hands of companies, the local authorities pay to the companies for water applied to public purposes, the sum paid forming part of the local taxation.

EXAMPLES OF RATES CHARGED, PER ANNUM, FOR DOMESTIC SUPPLIES. (*See page* 94.)

Name of Town or District.	Authority supplying the Water.	Basis of Charge.	Extra Charges for single Water-Closets, Baths, and High Service.	Annual Charges for Houses of the undermentioned Rents, or Rateable Values.					
				£10	£20	£30	£50	£80	£150
				s. d.	£ s. d.	£ s. d.	£ s. d.	£ s. d.	£ s. d.
Aberdeen	Corporation	Rent		2 6	0 5 0	0 7 6	0 12 6	1 0 0	1 17 6
		Public	Water Rate	2 6	0 5 0	0 7 6	0 12 6	1 0 0	1 17 6
Ashton-under-Lyne	Do.	Rent	W. C. 10/- each	9 0	1 2 6	1 12 6	2 10 0	3 5 0	...
Birkenhead	Do.	Annual Value		10 8	1 1 0	1 9 4	2 9 0	3 7 4	5 5 0
Birmingham	Do.	Rent		10 4	1 18 0	2 9 6	3 0 0	4 16 0	9 0 0
Blackburn	Do.	Rateable Value		16 0	1 12 0	2 8 0	4 0 0	6 8 0	12 0 0
Bolton	Do.	Rent		12 8	1 4 0	1 10 0	2 10 0	3 10 0	6 0 0
Do., to out Townships		Do.		13 0	1 6 0	1 16 0	2 16 0	4 8 0	8 5 0
Do.			W. C.	5 0	0 5 0	0 5 0	0 5 0	0 5 0	0 5 0
Do.			Bath	5 0	0 5 0	0 5 0	0 5 0	0 5 0	0 5 0
Bradford	Do.	Do.		15 0	1 10 0	2 2 0	3 5 0	4 16 0	7 10 0
			W. C.	3 0	0 4 0	0 6 0	0 8 0	0 8 0	0 12 0
			Bath	3 0	0 4 0	0 6 0	0 8 0	0 8 0	0 12 0
Brighton	Do.	Rateable Value		7 6	0 15 0	1 2 6	1 17 6	3 0 0	5 12 6
Bury	Do.	Rack Rental		15 0	1 10 0	2 2 0	3 5 0	4 9 0	7 10 0
			W. C.	4 0	0 5 0	0 6 0	0 6 0	0 7 4	0 10 0
			Bath	4 0	0 5 0	0 6 0	0 6 0	0 7 4	0 10 0
Carlisle	Local Board	Rent		9 6	0 14 6	0 19 6	1 9 6	2 4 6	3 19 6
			W. C.	7 0	0 10 0	0 10 0	0 10 0	0 10 0	0 10 0
			Bath	5 0	0 5 0	0 5 0	0 5 0	0 5 0	0 5 0
Dundee	Commission			13 4	1 6 8	2 0 0	3 6 8	5 6 8	10 0 0
Do.				Public	Water	Rate, 1d.	per £.		
Edinburgh	Trust	Rent		8 4	0 16 8	1 5 0	2 1 8	3 6 8	6 5 0
Glasgow	Corporation	Full Annual Value		6 8	0 13 4	1 0 0	1 13 4	2 13 4	5 0 0
Do.				Public	Water	Rate, 1d.	in £.		
Do., outside Borough	Do.			[illegible]	0 18 4	1 7 6	2 5 10	[illegible]	6 17 6

London	Chelsea Co., Grand Junction Co., New River Co., & West Middlesex Co.	Gross Annual Value		8 0	0 16 0	1 4 0	2 0 0	3 4 0	6 0 0
			High Service	...	...	...	0 4 0	0 6 0	0 8 0
			W. C.	...	...	...	0 4 0	0 6 0	0 8 0
			Bath	...	...	...	0 4 0	0 6 0	0 8 0
Do.	East London Company.	Do.		10 0	1 0 0	1 10 0	2 10 0	4 0 0	7 10 0
			High Service	2 6	0 5 0	0 7 6	0 12 6	1 0 0	1 17 6
			W. C.	...	...	...	0 4 0	0 6 0	0 8 0
			Bath	...	...	...	0 4 0	0 6 0	0 8 0
Do.	Southwark and Vauxhall Company.	Do.		10 0	1 0 0	1 10 0	2 10 0	4 0 0	7 10 0
			High Service	...	...	...	0 4 0	0 6 0	0 8 0
			W. C.	...	...	...	0 4 0	0 6 0	0 8 0
			Bath	...	...	...	0 4 0	0 6 0	0 8 0
Do.	Lambeth Company.	Do.		15 0	1 10 0	2 2 0	3 5 0	4 16 0	7 10 0
			W. C.	...	...	0 10 0	0 12 0	0 15 0	1 0 0
			Bath	10 0	0 10 0	0 10 0	0 10 0	0 10 0	0 10 0
Lowestoft	Company	Rent		12 0	1 4 0	1 16 0	2 15 0	4 8 0	7 10 0
			W. C.	10 0	0 10 0	0 10 0	0 12 0	0 12 0	0 12 0
			Bath	10 0	0 10 0	0 10 0	0 12 0	0 12 0	0 12 0
Manchester	Corporation	Poor Rate assessment		7 6	0 15 0	1 2 6	1 17 6	3 0 0	5 12 6
Do.				Public Water Rate, 3d. in £.					
Do., outside City limits	Do.	Rack Rent		10 0	1 0 0	1 10 0	2 10 0	4 0 0	7 10 0
Nottingham	Do.	Do.		11 0	1 0 0	1 10 0	2 2 0	2 14 0	4 10 0
Do., out townships, ¼th more than inside.									
Do.			W. C.	4 0	0 6 0	0 8 0	0 10 0	0 10 0	0 12 0
Do.			Bath	4 0	0 6 0	0 8 0	0 10 0	0 10 0	0 12 0
Perth, in original area	Do.			6 8	0 13 4	1 0 0	1 13 4	2 13 4	5 0 0
Do.				Public Water Rate, 1d. in £.					
Do., outside original limits				10 0	1 0 0	1 10 0	2 10 0	4 0 0	7 10 0
Plymouth, in Borough	Do.	Rent		8 0	0 14 0	1 0 0	1 8 0	1 16 0	2 0 0
Do., outside Borough,	50 % added.								
Preston	Corporation	Rateable Value		11 8	1 0 0	1 8 4	2 5 0	3 2 1	3 7 1
Sheffield	Company	Rent		15 0	1 5 0	1 17 6	2 16 3	3 15 0	5 12 6
			W. C.	6 3	0 6 3	0 7 6	0 9 5	0 9 5	0 12 6
			Bath	15 0	1 0 0	1 0 0	1 5 0	1 10 0	1 10 0
Southport	Do.	Rack Rent		11 0	1 2 0	1 13 0	2 15 0	4 8 0	7 10 0
			W. C.	4 0	0 6 0	0 8 0	0 10 0	0 12 0	...
			Bath	4 0	0 6 0	0 8 0	0 10 0	0 12 0	...
St. Helen's	Corporation	Gross Rent		9 0	0 18 0	1 7 0	2 5 0	3 12 0	6 15 0
Tranmere	Birkenhead Corporation	Rateable Value		10 0	1 0 0	1 10 0	2 10 0	4 0 0	7 10 0
Wirrall	Company	Rent		14 6	1 8 0	2 2 0	3 8 0	5 8 0	9 0 0
			Bath	3 6	0 3 6	0 4 6	0 6 6	0 7 6	0 12 0
Wolverhampton	Corporation	Poor Rate assessment		13 0	1 5 0	1 18 0	3 2 0	5 0 0	9 8 0
Do., outside Borough		Rental		14 0	1 8 0	2 2 0	3 10 0	5 12 0	10 10 0
Do.			W. C.	10 0	0 10 0	0 10 0	0 10 0	0 10 0	0 10 0
Worcester	Do.	Net Rateable Value		4 2	0 8 4	0 12 6	1 0 10	1 13 4	3 2 6
Do., outside City				10 0	1 0 0	1 10 0	2 10 0	4 0 0	7 10 0

The amount paid to a company by a local authority, for sanitary supplies, is generally made subject to agreement, but there are some cases in which the terms have been fixed by Act of Parliament. The following are examples :—

Date of Act.	Town.	Amount
1847	Leicester	Not exceeding one-half amount charged to any private consumer
1853	Bury	Not exceeding 4*d*. per 1,000 galls.
1854	Nottingham.........	Lowest rate charged to any private consumer
1854	Southport........ ...	Not exceeding 5*d*. per 1,000 galls.
1861	Wolverhampton ...	£400 per annum. (If profits of undertaking enable company to declare div. of 6 per cent., charge to cease.)
1860	Brompton District	At 1½*d*. per ton, but if hydrants fixed and repaired by company, charge to be 2*d*. per ton. Equal to about 6*d*. and 9*d*. per 1,000 galls.
1860	Maidstone	
1861	Northampton	
1865	Gainsborough	4*d*. per 1,000 galls. Local Board to pay expense of pipes, hydrants, &c. Subject to slight reduction when company declares higher dividend than 5½ per cent.
1868	Slough	Not exceeding 6*d*. per 1,000 galls.
1868	Windsor and Eton	
1869	Harrowgate.........	9*d*. per 1,000 galls.
1869	Bishop's Stortford.	Flushing free. Watering streets £50 per annum.
1870	Chiltern Hills	Not exceeding 1*s*. per 1,000 galls.
1870	Littlehampton.....	Not exceeding 6*d*. per 1,000 galls.
1859	Norwich	Up to 7,000,000 galls., at 7*d*. per 1,000 galls. Over 7,000,000 galls., at 6*d*. per 1,000 galls. For cleansing purposes, to 50,000 galls. at 10*d*. Over 50,000 galls., at 1*s*. per 1,000 galls.

In some of these towns the waterworks are now in the possession of the public authorities.

CHAPTER VII.

WATER APPLIANCES FOR DOMESTIC SUPPLIES.

Control over pipes and fittings—Construction and management of cisterns—Overflow-pipes—Examples of defective and dangerous work—Water direct from mains—Water-closets—Faulty closets—Entrance of sewer gases—Drawing-cocks—Stop-cocks—Ball-cocks—Iron and lead pipes—How to lay new pipes—How to discover underground leakages—Pipes in bad soil—Baths—Hot water apparatus—Outside stop-cocks—How to choose water fittings—Waste-preventers—No water—Water-traps and their defects—Experiments on passage of gases through water—Effect of frost—How to thaw frozen pipes.

THE first step to be taken in order to insure an effective control over the sanitary and water appliances of a house or other building is to obtain by a personal examination, assisted if necessary by a practical plumber, a thorough knowledge of the position and state of every pipe, cistern, tap, and trap, inside and outside of, or in any way connected with, the premises. The information thus obtained should be marked upon a plan.

Having got this preliminary knowledge, the directions given in the following pages can be intelligently carried out, and the principles laid down properly applied.

Let it always be remembered that upon due attention to these matters, the health and the life of every member of a household may depend.

CISTERNS.

As to construction.—Cisterns to store water for potable purposes should be of slate, iron, stone, glass, or brick with a lining of Portland cement. The use of lead-lined cisterns for storing drinking-water should be avoided. Although water seldom acts on tarnished lead, there are some conditions favourable to its action, and on the whole it is better not to employ this material for cisterns from which water is to be drawn for dietetic purposes. If existing lead cisterns are repaired or relined, the new bright lead should be exposed to water for a few days before the cisterns are again brought into use. Water which does not affect tarnished lead often acts violently upon bright lead. Zinc is also acted on by water, and may produce metallic poisoning if used for cisterns. Timber cisterns without any lining, and galvanized iron cisterns, are objectionable, and they are not durable.

If iron cisterns are used, they should receive a coat of boiled linseed oil, carefully applied, before they leave the ironfounders' premises, and before they are painted A wash of Portland cement affords an excellent protection for iron, if skilfully put on. The iron should receive three coats, and the wash should be renewed every year. Every cistern should have a substantial, well-fitting lid. Neglect of this provision is a prolific source of discomfort and disease. Any noxious gases which may come into contact with the water are very readily absorbed by it. Floating particles of organic matter, and other objectionable and deleterious substances with which the atmosphere of towns is charged, enter the cisterns; very frequently animals, such as cats, rats, mice, and birds, fall in, and remain there until they reach an advanced stage of decomposition. When the water becomes offensive, a plumber is sent for or complaint is made to the water officials, and when an examination is made, the cause of the complaint is discovered. Cisterns should be easy of access. As a rule they are exceedingly difficult of access, and in other respects are placed in most unsuitable positions; and they are treated as mysteries which only plumbers can be expected to understand. It often happens that their very existence is unknown to the tenants of the houses to which they belong. All this is the very reverse of what it ought to be. Cisterns should be placed where they can be seen and examined without difficulty. If fixed in an upper room, they should not (unless well covered) be put under a skylight. The cistern-room should be well ventilated. In every household there should be a periodical inspection and cleaning of the cisterns.

One of the greatest evils in connection with cisterns is the dangerous and disgusting practice of joining overflow pipes to water-closet drain-pipes. These overflow pipes frequently form channels of communication between the sewers and the cisterns; and the sewer gases pass freely to the water. The community has suffered enormously from deterioration of health caused by the use of water contaminated in this way by exhalations from drains and sewers.

The accompanying section represents a case that came under my observation, where all the members of a family had for many months suffered from serious illness, before the cause was discovered and remedied. A is a cistern with a cover firmly screwed on; B is the overflow-pipe, passing through the bottom of the cistern to the outlet of the water-closet siphon; C is a trap intended to seal the overflow-pipe, and prevent the ascent of sewer gas. When the cover was removed to examine the cistern, the stench was almost intolerable: the trap contained no water, and the overflow was therefore simply a communication-pipe between the sewers and the interior of the cistern. This illustration represents a large class of cases.

There is a still more numerous class of cases in which the overflows are laid to the trap of the siphon, or to the inlet side at D or E. These, though much less objectionable than the former, are equally to be condemned.

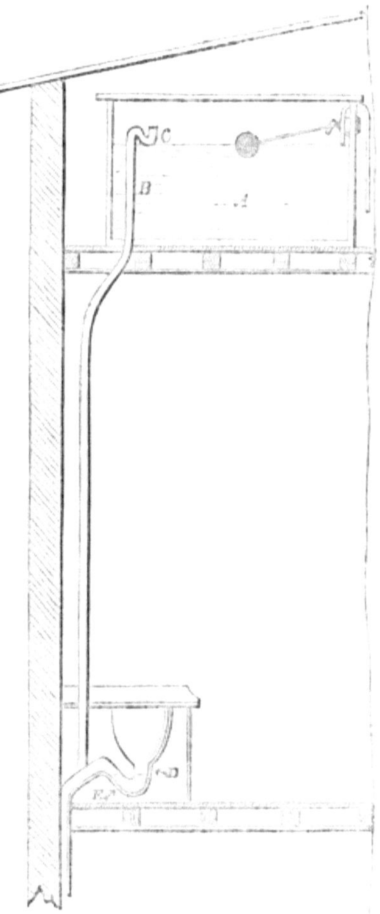

No connection should, upon any consideration, be allowed between cisterns and drain-pipes or water-closet flush-pipes; wherever such connections exist, they should at once be cut off.

The following illustration represents another class of cases, where danger arises from allowing water-closets to be supplied from cisterns which also supply water for drinking.

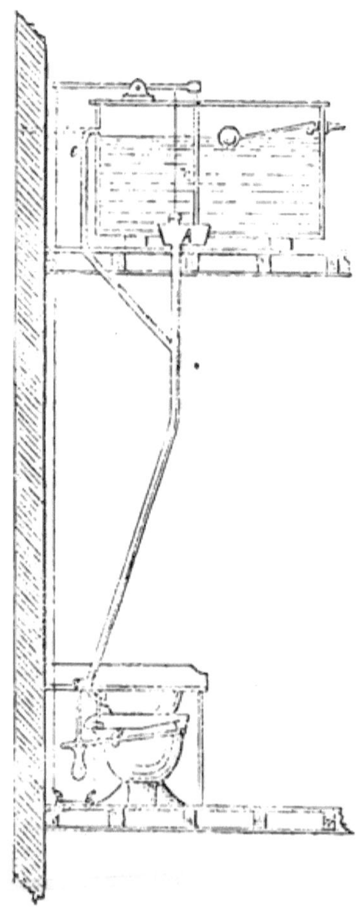

A is a small lead service-box, fixed in a large cistern to supply an after-flush to the pan-closet; B is an air-pipe leading from the service-box to a point in the cistern over the top water line. The illustration represents an instance which came under my notice, in which the water used for drinking was polluted by foul gases passing through the air-pipe. These air-pipes are very frequently used as waste-pipes. In the drawing I have shown a separate waste-pipe, branched into the flushing-pipe. The offensive character of the above arrangement will be obvious to every one. The dotted lines show what alteration is required. The condemned connections to be cut off. The air-pipe to be carried through the cover of the cistern and through the roof of the house. The overflow or waste-pipe to be laid to the outside of the building. But better still would it be not to permit water-closets to be supplied, under any circumstances, from receptacles for potable water.

The next illustration is taken from an American sanitary journal, and represents a cistern in a well-known New York hospital. The cistern was constructed of wrought iron, the overflow-pipe was branched into the soil-pipe, and the joint so badly made that the sewer gases escaped from it, as well as having free course into the cistern: the air-pipe was added because a difficulty was experienced in getting water down the draw-off-pipe. When this disgraceful state of things was discovered

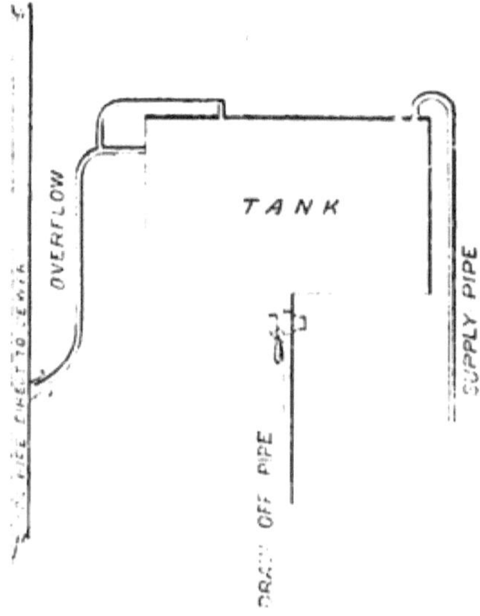

there were fifteen children sick with scarlet fever in an adjoining room! There is a combination of blundering about this piece of work that can only be fully appreciated by a mechanic.

Where cisterns are placed in an exposed position, and therefore liable to be affected by frost, they should be carefully protected from the influence of the weather by an exterior lining of brickwork or felt. Sawdust held in by timber may be used, if care is taken to preserve it from wet.

TAPS IN DIRECT COMMUNICATION WITH THE MAIN.

Wherever water is distributed under the constant supply system cisterns are not necessary, except for water-closets and hot water apparatus, or to provide a reserve when water is turned off from the mains for repairs. Where cisterns are used there should always be a tap attached to a pipe leading direct from the main. By this means the water is obtained in the best possible condition. If the water is temporarily discoloured, owing to the disturbance of iron rust in the mains, the tap should be left open until the water becomes clear before drawing any for use.

WATER-CLOSETS.

There are so many different kinds of water-closets made that it would be impossible within the limits of this book to describe them and discuss their merits and defects. The principal points to be observed in the construction of a water-closet are :—

1. The form of the basin should be such as to prevent as far as possible the accumulation of filth on its sides. The basins known as "long-hoppers" do not comply with this condition, and are therefore objectionable. The force of the objection will be apparent on comparing the following sketches :—

2. The arrangement for flushing the basin and removing its contents should be such as to effectually cleanse the basin and carry the soil to the sewer. To accomplish this the flushing-pipe should not be of less diameter than 1¼ inch (internal), and it should be fed from a flushing-cistern capable of giving at least two gallons at each flush. The distance from the closet seat to the underside of the cistern should be not less than 4 feet 6 inches. The manner in which the flushing-pipe is connected to the closet-basin has a great deal to do with the efficiency of the flushing; but this is a matter of too technical a character to be dealt with here.

3. There should be no possibility of sewer gases, or other filth, collecting between the basin and the trap, as in pan-closets. The pan-closet is one of the most objectionable forms in use.

4. The trap should be effective, and not allow sewer gases to escape when the basin is flushed. On this point see remarks on traps, at page 129.

5. The soil-pipes should be carried up, without any diminution of diameter, to the outside of the premises above the roof, and at a safe distance from any window.

6. The arrangement for supplying water to the closet should render impossible any contamination of the water used for drinking or culinary purposes. This contamination may be communicated in several ways, and in order to determine whether there is danger in any given case, attention must be paid to the following facts:—

(a.) That sewer gases will pass through water; and that a water-trap, if not deep enough and its contents frequently renewed, is not a sufficient protection.

(b.) That when the supply-pipes are empty, in consequence of the water being turned off at the main, a partial vacuum is formed; and if the apparatus by which the water closet is supplied is defective, or is so fixed and constructed as to leave an open passage between the closet and the street water-main, noxious gases, or any foul stuff that may be in the closet-basin, will be forced by the atmosphere into the empty water-pipes, to mix with the water for general distribution, when the supply is again turned on.

The example given here represents a form of water-closet which is, unfortunately, very common in some towns. A is the leading pipe from the main; B the communication-pipe leading to the water-closet, and sometimes actually dipping into it, as indicated by the dotted lines; c a stop-cock by which the supply of water is controlled. If the street-main is shut off, say for the insertion of a ferrule by the waterworks men, a partial

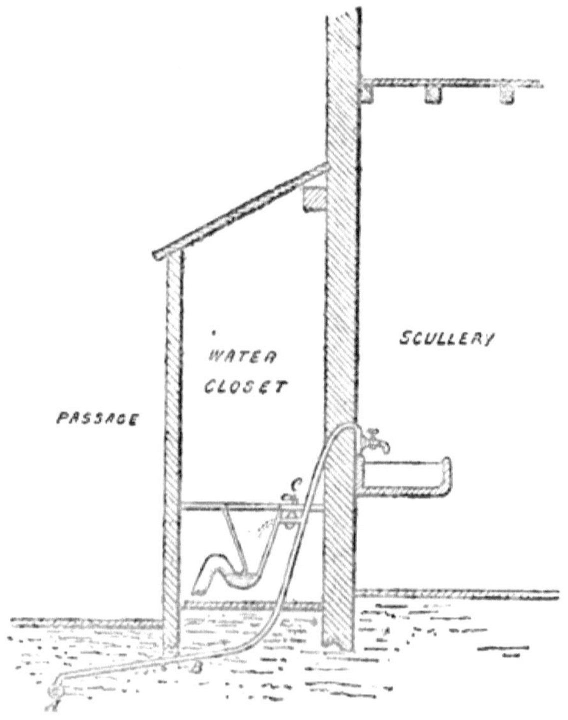

vacuum takes place in A and B: let the stop-cock C now be opened, and B becomes a channel through which any filth, liquid, or gases that may be in the closet-basin are greedily taken into the main. Those who have had opportunities to become acquainted with the habits of the lower classes know how frequently their closet-basins are full of filth. If instead of being in direct communication with the street-main, the closet is fed from a storage cistern, the same effect may, to a limited extent, take place if the cistern is at any time empty.

People sometimes imagine that a small flow of water trickling continually down a closet promotes cleanliness and prevents smells; this is a delusion. The water is simply wasted. Others make a practice of fastening the handle of their closet occasionally, so as to allow water to flow through, "to keep away bad smells;" this is an equally mistaken and more objectionable practice, and those who indulge in it render themselves liable to a penalty of £5 for wasting water. What is wanted effectually to flush a closet soil-pipe is to let down quickly a sufficient volume of water to carry before it every impurity. An anecdote is told of a domestic servant from Glasgow who, having taken a situation in another town, went to her mistress when she was about to leave for a summer holiday, and asked, "Please 'm, shall I fasten the handle up?" The lady did not understand the question. In explanation, the girl said that in Glasgow, when shutting up the house for summer holidays, it was always the custom to prop up the handle of the closet, "to keep the drains sweet," until the family returned. If this story is true it explains the extravagant consumption of water in that city.

BALL-TAPS.

There are numerous kinds of ball-taps which are almost equally simple, strong, and easily repaired. The sketch represents one of these. The ball is made of

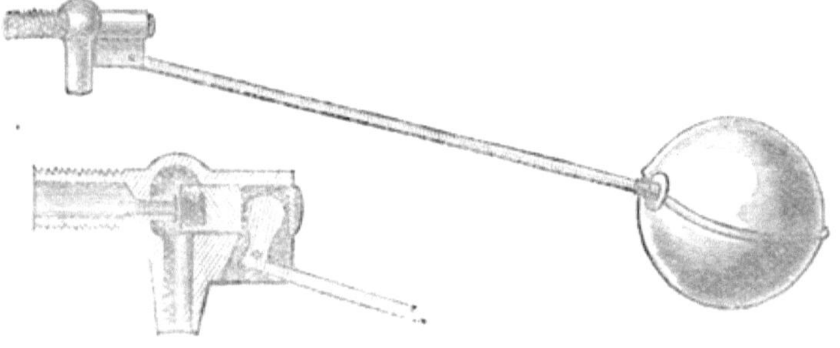

copper in two halves, which are united by solder. Some waters act violently upon solder, and the balls are soon water-logged. In such cases the balls should be coated at the joining of the two halves with knotting, or, which is better, a float may be made of a different form, with no solder joint near the water-line, and with the lever fastened to the top of the float, instead of to the centre.

REGULATING CISTERNS FOR WATER-CLOSETS AND URINALS.

The simplest and most effective form of regulating apparatus at present in use for flushing water-closets and urinals is that known as the Double-valve Cistern.

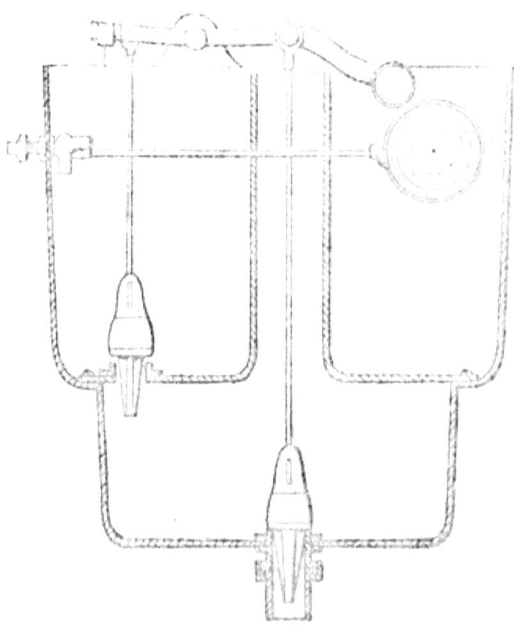

The principle on which this cistern is constructed is that of preventing water flowing in and out at the same time, thus making it impossible to send a continuous run through the closet. As will be seen from the sketch, there are two compartments connected by an opening on which is placed a valve controlled by a lever, to one end of which the pull is attached, while, suspended from the same lever, and descending to the lower division, is a second valve, which governs the outlet to the closet flushing-pipe. When the handle is pulled the upper valve falls on to its seat, the outlet-valve is raised, and the contents of the flushing chamber discharged. No water can pass from the upper part into the lower when the outlet-valve is open, and a second flush cannot be obtained until the handle has been released, when the outlet-valve descends to its seat and the inlet-valve is opened. The water then passes freely into the flushing-chamber, and in doing so allows the ball to fall so that the cock opens and the cistern is again filled.

The valve washers are made of india-rubber or leather (I prefer leather), and are easily renewed. The valves should be regulated so that the outlet cannot be raised until the inlet-valve has been safely seated.

TAPS.

Taps used for controlling the discharge of water from a pipe, or direct from a cistern or boiler, are called drawing, or draw-off taps, or cocks. Taps fixed on a line of pipe, to regulate and control the flow through it, are termed stop-taps, or stop-cocks. Taps actuated by a ball, or other float, are called ball-taps or ball-cocks.

Classifying draw-off and stop-taps according to the principles involved in their construction, they may be divided into plug-taps, screw-down taps, and self-closing taps.

Plug-Taps.—The objections to a plug-tap are two:—

1. It closes so quickly that a strain is put upon the pipe, which gradually weakens, and finally bursts it by the sudden stoppage of the flow.

Water is a non-elastic body, and when a column flowing through a pipe is suddenly arrested, a concussion takes place which is often heard and felt for a considerable distance. The greater the velocity of the water in the pipe the more violent will be the rebound.

2. A defective plug-tap cannot be repaired without employing a mechanic. If a plumber is called in to grind the plug of a tap, his charge will probably amount

to the cost of a new tap. Plug-taps are useful as drawing-off taps from boilers, and as cistern stop-taps.

SELF-CLOSING OR SPRING-TAPS.—These taps have the same defect as plug-taps in arresting the water too suddenly, and thus injuring the pipes to which they are connected. Air-vessels are usually attached to pipes on which self-closing taps are fixed, so as to lessen or prevent the concussion, but they are seldom effective.

SCREW-DOWN TAPS.—A screw-down tap closes slowly. The velocity of the water is gradually retarded, and the

flow is finally stopped without producing any concussive strain on the pipe. It is very easily repaired, and, if made of good metal, is very durable. The washer can be changed by any person of ordinary intelligence. For cold water, the washers are made of leather. For hot water, specially prepared washers can be bought from any dealer in taps. It frequently happens that persons who have been accustomed to use plug-taps do not understand how to open and close screw-down taps. A Local

Board engineer recently had occasion to call at a cottage in his district, and was asked to look at the tap. "It has been here," said the woman, "about a year, and scarcely any water will come out of it, and it's always leaking." The tap proved to be one of the screw-down kind, and the cause of the leakage was simply that it had never been properly opened or shut. It had been treated as a plug-tap.

Although considerable waste takes place from the imperfect shutting of screw-down taps, experience shows that the total waste of water from a number of screw-down taps is less than from an equal number of plug-taps.

Servants should be warned never to leave a tap open if on turning it they find that there is no water in the pipe. Carelessness on this point frequently leads to considerable damage to property by flooding. In towns supplied on the intermittent system it is a very common practice to leave taps open until the water is turned on. This practice occasions great waste.

PIPES.

There is a great deal of prejudice against the use of lead water-pipes. This prejudice arises from the belief that the action of water, especially soft water, upon the lead may cause lead-poisoning. Soft waters do not necessarily act upon lead. Waters that act violently on bright lead act only slightly or not at all on tarnished lead, and as lead quickly becomes tarnished on exposure to the atmosphere, there is practically little risk in using lead pipes. After water flows for a short time through a pipe the interior is coated with a film which prevents any injurious effect. The chief objection to wrought-iron pipes is, that with soft water they are choked by rust in a few years. Where there is any apprehension of lead-poisoning, iron pipes coated by Dr. Angus Smith's process, or treated by Professor Barff's method, or lead pipe with a lining of tin may be employed. In ordering a tin-lined pipe, care must be taken that the tin forms an independent tube within the lead, and that the outer casing of lead is of sufficient thickness to resist the maximum internal pressure which the pipe will have to bear, independently of the tin-lining, also that the tin is of uniform thickness throughout the pipe. Unless these conditions are observed, the tin-encased lead pipe will prove a most unsatisfactory investment.

Galvanized and enamelled iron pipes I cannot recommend, for conveying soft water.

When pipes are being laid in a house for the first time,

every one who has any authority over the work, whether architect, owner, or tenant, should resolutely prohibit the dangerous and costly practice of embedding the pipes in the plaster-work of the walls, or any other form of covering which makes it difficult to get at the pipes for examination and repairs. All internal pipes, both soil-pipes and water-pipes, should either be exposed and neatly painted, or laid in timber casings, constructed in such a manner that they can be easily opened.

In spite of the numerous instances in the experience of every plumber, in which property is destroyed and life endangered in consequence of pipes being laid behind plaster-work and in inaccessible places, architects continue to specify work in this objectionable way, and they will make no change until compelled to do so by those who employ them.

Lead pipes should not be laid in ground containing cinders or chemical refuse from buildings, or lime in any form. If pipes are so laid, they will probably require renewing in two or three years.

Where it becomes unavoidable to lay pipes through ground of the kind described, they should be laid in a wooden trough, V shaped, and covered with asphalte. Troughs of this shape, specially made for the purpose, can be had. Failing this, the method of laying may be adopted as shown on page 121.

UNDERGROUND LEAKAGES.

Recent investigations with respect to waste of water have shown that there are immense numbers of leakages from private pipes under houses, yards, gardens, and passages, the existence of which leakages is not known and is not suspected by the occupants of the premises. Such leakages not only endanger the foundations of the houses, but they imperil the lives of the inhabitants by making the subsoil damp and unhealthy.

Householders can generally ascertain if there are defective underground pipes about their premises by the following simple proceeding:—Shut off all the drawing-taps in the house, and tie up the cistern ball-cocks so that no water can escape from them. Take a wooden rod —nothing is better than a Malacca cane—or a common poker, and place one end on an exposed part of the pipe or on a tap, and apply the other end to the ear. Let this be done in various parts of the house. If there is any defect, a peculiar hissing sound will be heard. If any difficulty is experienced in distinguishing this sound, familiarity with it may be acquired by slightly opening a tap near a point where the rod is applied. Care must be taken that all the stop-cocks are open.

PRIVATE BATHS.

All arrangements such as three-way cocks, by which hot and cold water are brought in and waste water is taken out by the same passage, are objectionable. Ordinary screw-down valves answer every requirement for filling baths. The inlet should always be at the top of the bath, and entirely separate from, and unconnected with, the outlet. For the outlet a well-ground brass plug is simple and effective. The common practice of connecting waste-pipes from baths to water-closet traps or soil-pipes is very objectionable.

HOT WATER APPARATUS.

The subject of hot water apparatus is of too technical a character to be dealt with in a work intended only for popular use. As in works on health readers are advised if certain symptoms appear to consult a physician, so in reference to heating apparatus my advice is, if there is anything wrong, consult an engineer or plumber. But there are one or two points in regard to which some information may be given. First, as to the principle on which heating appliances in private houses are arranged. The water is generally heated in a boiler at the back of the kitchen fireplace. If the hot water is distributed through pipes, the boiler is fed from a cistern at the top of the house. This cistern is fixed near to the principal storage cistern, and is supplied from it through a pipe, on which a check-valve is placed to prevent the hot water mixing with the cold. From the hot water cistern a pipe is brought down to near the bottom of the boiler, and from the top of the boiler a pipe ascends to the cistern. In this way a constant circulation is maintained. The hot water supplies are taken from the ascending pipe. Occasionally the contents of the hot and cold water cisterns get mixed, and tepid water is served instead of cold. When this occurs the check-valve will probably be found defective. How can the bursting of domestic boilers be prevented? This is a question often asked. Perhaps it may be as well to explain, for the information of those who are ignorant of this elementary fact, that the danger of a boiler bursting arises either from an obstruction to the circulation or from the introduction of cold water into a hot empty boiler. The circulation may be—and often is—impeded by the freezing of the water in the pipes during a severe winter. The steam or hot water cannot escape, and the boiler bursts. This may be prevented by the application of a small safety-valve or by a relief-pipe.

It would be an useful addition to every domestic boiler if a glass gauge were placed in front of the fireplace to show the depth of water in the boiler. Allusion has already been made to the incrustation and consequent bursting of boilers in districts supplied with hard water. This may to a great extent be obviated by frequently emptying and washing out the boilers, carefully removing the scale found on the bottom and sides.

STOP-TAPS.

Every house ought to be provided with a stop-tap, either outside of or immediately within the curtilage of

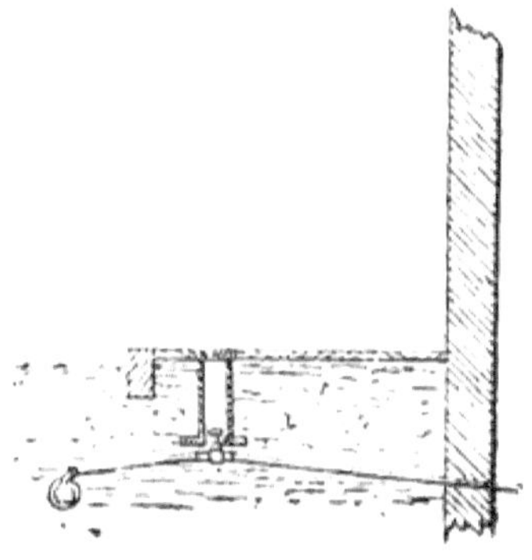

the property, so that the water may be shut off during repairs, and when the house is unoccupied, and every night during frost. It is also desirable to place a stop-tap on every pipe leading from a cistern, so that repairs may be effected on that particular line of pipe without emptying the cistern.

QUALITY OF WATER FITTINGS.

The system of testing and stamping fittings, which has been established in Manchester, Liverpool, and other towns, by the waterworks authorites, gives to the inhabitants of those towns a sufficient guarantee for the quality of the fittings they buy. Where there is no official inspection, purchasers will find their best protection in dealing with manufacturers of established reputation, and in aiming to secure, in the articles they require, simplicity, durability, and ease of repair. All cocks should be made of hard brass or gun-metal.

WASTE-PREVENTERS.

The last few years have witnessed the invention of a multitude of appliances, many of them very ingenious, known as waste-preventers. They may all be included in one of two classes. 1. Restrictions in communication-pipes, such as a metal disc containing a small hole, designed to prevent more than a given volume of water passing through a pipe into a house in any one day. 2. Taps so contrived that on being opened they allow only one or two gallons to flow through, and then close.

There may be exceptional cases in which some of these contrivances may prove useful, but their general use is certainly not advisable, and not necessary. They are chiefly to be met with in London, and I believe that their introduction is due to engineers whose experience is confined to a wasteful intermittent service, and who imagine that if a constant supply is given through ordinary taps, a large proportion of them will always be left open.

NO WATER.

This is a very common form of complaint to water-works officers. "There is no water at No. — Blank Street." When the complaint is investigated, the want of water is traced, in the majority of instances, to some defect in the house appliances. A ball-cock is disarranged, or a stop-cock has been inadvertently closed, or a waste is taking place which prevents the water from rising to the cistern. It would therefore be well, before making a complaint, to ascertain that the ball-cock is working properly, the stop-cock open, and that the deficiency experienced at one point is not due to waste at another.

Defective ball-cocks and neglected cisterns explain most of the complaints that reach water offices with respect to "no water" and "dirty water."

WATER-TRAPS.

Reference has already been made to traps connected with water-closets, baths, and other appliances. The subject is rather one of house drainage than of water supply, and I shall confine my remarks to that branch in which the agency of water is relied upon as a seal. Water is almost universally used, in preference to any mechanical contrivance, to prevent the return of noxious gases through waste-pipes, overflow-pipes, and soil-pipes, that are connected with drains and sewers. There are endless varieties of traps, but they are all based on the principle of a tube bent so as to retain water. In some houses there are several dozens of these water barriers, upon the efficiency of every separate one of which the occupants of the house depend for protection against sewer poisoning. In the construction and management of water-traps there are three facts to be always kept in view:—

1. That sewer gases will pass through water.

Some interesting experiments on the passage of gases through traps were made a few years ago by Dr. A. Fergus of Glasgow. At the outlet end of a trap (a bent tube) he placed a small vessel containing the test solutions A, and at B the test papers were suspended. He found that ammonia passed through the water in from fifteen to thirty minutes; sulphurous acid in an hour; sulphuretted hydrogen in three to four hours; chlorine in four hours; carbonic acid in three hours. He afterwards fixed a ventilating pipe, as indicated by the dotted lines, and obtained similar results, but the reaction was a little slower in showing itself.

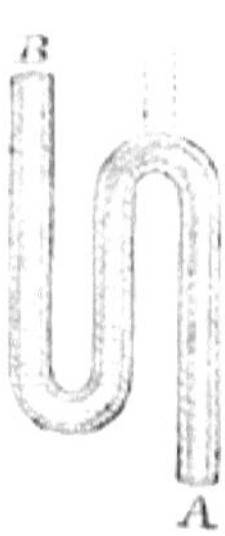

2. That traps may be emptied by evaporation. If traps are placed where the water is not frequently renewed, or if a house is long unoccupied, danger may arise from this cause.

3. That if two or more traps are connected with the same line of pipe the flushing of one may empty the

others. There have been cases in which foul smells from wash-basins and sinks have been long complained of, and considerable expense has been incurred in seeking remedies, before the cause has been traced to the emptying of the traps by siphoning. To prevent this action, every trap should be properly ventilated. Waste-pipes from baths, lavatory-basins, butlers' pantries and housemaids' sinks should be brought to the outside separately, and not joined to soil-pipes.

EFFECTS OF FROST.

In England it is seldom that frost penetrates to a greater depth than $2\frac{1}{2}$ feet below the surface of the ground, and it is the practice of waterworks engineers to lay their pipes at about that depth. This applies only to public pipes, but in many towns the same rule now extends to private pipes laid underground, and no pipes or fittings are allowed to be laid above ground without adequate protection against frost. To protect pipes in exposed situations a covering of felt or sawdust, or other non-conducting material, must be employed. The protection of cisterns is referred to at page 108. The bursting of cisterns may be prevented by putting in a block of wood, or a weighted hollow india-rubber ball, or other elastic body. If pipes are laid against an external wall they should be fastened to a board, and not allowed to come into direct contact with the wall.

The poorer classes are the first and chief sufferers from the freezing of water in pipes. Outside-cocks, uncovered pipes in courts and yards, soon feel the effects of a low temperature. Taps and pipes of inferior quality are less capable of resisting the attacks of frost than strong pipes and well-made taps. With a continuance of frost every class suffers, and explosions of kitchen boilers are frequent. When a thaw sets in the ice-bound pipes become free, and wherever a hole has been made by the expansion in freezing a leakage takes place. Houses are flooded, water-offices besieged, and plumbers reap a rich harvest.

What can be done to prevent freezing in pipes that are unavoidably exposed? The common practice is to leave a tap slightly open, so as to maintain a constant current through the pipe. This plan is wasteful, and is not always successful. Perhaps the safest course is to empty the pipes and cisterns, and only to allow water to flow in from the main as it is wanted for consumption. To do this an outside stop-cock is required on the service-pipe, and a drawing off-cock at the lowest point in the course of the pipe inside the building. It also requires more intelligence and attention than domestic servants usually display.

To thaw a frozen pipe, the simplest and safest way is

to pour hot water upon it, or apply cloths dipped in hot water to those points where the pipe is most exposed. The freezing will generally be found to have taken place near a window, or near the eaves of the roof, or at a bend.

If pipes are frozen and a thaw is expected, care should be taken to close all stop-cocks as a precaution against flooding. To prevent kitchen boilers exploding, it is necessary to see that they always contain water, and that there is no stoppage in the pipes connected with them.

CHAPTER VIII.

WASTE OF WATER.

Significance of waste—Extent and value of waste from defective fittings—Relation of pressure to waste—Standards of consumption—Waste in the metropolis—Wasteful constant supplies—Experience of American towns—Results of careless methods—Sources of waste—Prevention of waste—Statutory powers to prescribe fittings—Reductions effected by proper regulations—Liverpool system—Co-operation of public desired—Responsibility for waste prevention.

THERE is no part of a waterworks engineer's duty so disagreeable as that connected with the prevention of waste from private fittings. As a general proposition, the importance of checking waste is readily admitted, especially if its prevalence threatens to necessitate the imposition of a tax for an additional supply; but the application of waste-preventing measures to individual cases is unpleasant and unpopular.

A little rill, trickling down a mountain-side, looks very insignificant; nevertheless, little rills make big rivers. So with waste: a slight leakage from a tap may seem very unimportant; but slight leakages gradually exhaust huge reservoirs. A mere dribble from a tap, continued for a year, represents in money value the water rate of an artizan's cottage. It is not easy to convince people of the extent and effect of waste in their own houses, and of the necessity of replacing fittings which are known to be of a wasteful character: hence, any attempt to interdict the use of such fittings excites strenuous opposition, and makes it difficult to obtain the requisite Parliamentary powers, or to enforce them when obtained. It is the difficulty of exercising a control over fittings that has led water authorities to resort so extensively to the obnoxious intermittent system, as a means of protecting themselves, by limiting the duration of the supply to the time in which the quantity they can afford to deliver is consumed.

The average amount of waste that takes place from various kinds of defective fittings can only be estimated with any pretence to accuracy by measuring a great number of individual leakages from each kind of fitting, and even then the results obtained will only be applicable to the town or district in which the observations are made, because of the diversity in pressure to which the fittings of different towns are subjected. It is seldom that measurements such as are here indicated are taken, or required, on a scale sufficiently extensive to give useful average results; but I have before me now a statement which shows the volume of waste from more than a thousand separate leakages in Liverpool, ascertained in each case by observing the number of seconds in which the leakage filled an imperial measure. The results are as follow :—

		Per hour.	
Average waste from	defective cocks	9¼	gallons
Do.	defective ball-cocks in store-cisterns	16½	„
Do.	defective pipes	30	„
Do.	defective water-closets, including all water-closet appliances.........	18½	„
Mean of the above...	....	17	„

These are the ordinary leakages discovered by waste water inspectors in the course of their domiciliary visits. The fittings are of an average quality, and have for many years been systematically inspected. A leakage of 17 gallons per hour, with a constant service, would afford a liberal domestic supply to forty persons.

The engineer of the Glasgow Waterworks, in 1860, found that, from badly constructed and leaky taps alone, the waste in that city was 7,200,000 gallons per day, being equal to 20 gallons per head of the entire population, and equal in money value to £50,000 per annum.

The head, or pressure, under which water is distributed is an important element in determining the amount of waste occasioned by defective fittings. The significance of this may be inferred from the following figures, showing the waste through defective cocks under varying heads.

	[illegible] in gallons per hour from same defect, under various pressures, in lbs. per square inch.	
	42 lbs.	[illegible] lbs.
A defective plug-tap	7·0	2·7
Another ditto	27·7	8·0
A defective screw-down tap	32·7	1[illegible]·7
Another ditto	18·0	6·0
Another ditto	18·0	2·5
Another ditto	2·6	0·4

These are wastes from individual taps, and they do not convey an adequate idea of the influence that the pressure under which water is distributed has in determining the total volume of waste that obtains over a large area. The following table shows the effect produced on the waste in a number of town districts by reducing the pressure of water in the distributing mains, by means of pressure reducing valves, from its normal state to a pressure which was amply sufficient for all the requirements of the inhabitants. The experiments were made between midnight and 1 a.m., when it was evident that nearly all the water flowing into the districts was being wasted.

Town District.	Number of Inhabitants.	Before reduction of pressure. Pressure in lbs. per sq. inch.	Before reduction of pressure. Approxmt. waste in galls. per hour.	After reduction of pressure. Pressure in lbs. per sq. inch.	After reduction of pressure. Approxmt. waste in galls. per hour.	Per centage of reduction.
A	3,950	70	4,200	28	2,200	48
B	5,500	52	2,200	29	1,300	40
C	2,720	70	1,700	38	880	49
D	3,325	65	750	30	360	52
E	5,738	45	1,330	25	960	27

The total amount of waste taking place in a town may be inferred from the tables on pages 89 to 91. As a measure of the quantity required to give an ample supply to a town population, Manchester and Liverpool may be taken as standards for comparison. In both places an unrestricted constant supply is distributed to the inhabitants. In Liverpool the shipping trade constitutes a special demand. There are also extensive sugar-houses, and numerous hydraulic apparatus; public baths and washhouses are provided on a larger scale than anywhere else in the country; private baths are common; water-closets are general; and many trade supplies are given without the intervention of meters. Under these conditions, so favourable to a large use of water, the consumption for domestic purposes, including hotels, offices, public houses, shops, stables, warehouses, and all waste, is under 15 gallons per head, and for trade and public sanitary purposes $7\frac{1}{2}$, making a total average consumption for all purposes of 22½ gallons per head per day. Even of this a large proportion is wasted, but it is waste which cannot be materially reduced without oppressive and costly proceedings. In Manchester, the manufacturing industries form a special demand, but the corporation have not adopted the water carriage system for sewage disposal, and water-closets are not numerous. The consumption for domestic purposes, including waste, is 13 gallons per head, and for trade purposes 7 gallons per head, making a total for all purposes of 20 gallons per head per day.

In the metropolis the total consumption with an imperfect intermittent supply is 32½ gallons per head per day, of which from 7 to 7½ gallons are for other than domestic purposes. There is no reason why the metropolis should not be supplied on the constant system with an average consumption of 22½ gallons per head per day without imposing burdensome regulations upon consumers. But, allowing 25 gallons for the London supply, a constant service might still be introduced with a saving in water of 7½ gallons per head upon the present intermittent service. Taking the population at 4,450,000, a reduction of 7½ gallons per head would amount to 33,375,000 gallons per day, which is sufficient to provide a liberal supply for a population of 1,335,000. The cost of pumping this water is £32,610 per annum, and the cost of filtration £3,650 per annum. At the average price charged to consumers the money value is £314,818 per annum. If its value is calculated at the estimated cost of obtaining additional supplies from the present sources, as recently given in evidence, it represents an outlay of £2,886,000. All these figures, it must be remembered, refer to the reduction that could be effected in the consumption by substituting a properly regulated constant supply. But the actual waste under the existing intermittent system is far in excess of this. Fully one-half of the water now distributed by the companies is absolutely wasted. Under a constant system the useful consumption of water would be greater than at present, while the waste would be much less. The evils of the intermittent system are so serious that a constant service would be desirable even though it involved a very much larger expenditure of water; but the fact that the improved system could be established with an economy in water is an additional argument in its favour.

The full effect of wasteful methods and appliances is only experienced in those places where an intermittent supply has been followed by a constant supply from sources largely in excess of immediate requirements. When works are carried out which are calculated to meet the growing wants of a town or district for a considerable number of years, it is almost impossible to induce people to adopt precautions and to submit to regulations of which the pressing necessity cannot be shown. The introduction of water from a new source yielding a supply by gravitation greatly beyond the immediate demand has in almost every instance led to carelessness on the part of waterworks managers, and extravagance on the part of consumers. The delivery of the Loch Katrine water into Glasgow affords an example of this. The consumption in Glasgow rose to such an alarming extent soon after the Loch Katrine water was brought into the city that attempts were made to check the waste, but the prevailing feeling that there was a practically inexhaustible lake to draw upon prevented the enforcement of suitable regulations. The present consumption in Glasgow is at the rate of 50 gallons per head per day.

There have been similar experiences in many American cities. In New York the engineer who designed the Croton Aqueduct estimated the requirements for a liberal supply 30 years ago at 25 gallons per head. The actual consumption at present is about 75 gallons per head. That most of this is wasted has been proved by observations on the discharge from the reservoirs at hours when very little water is being consumed for any useful purpose.

The volume of water flowing into New York from the Croton Aqueduct is enough to supply more than double the present population at the rate per head originally estimated, but the consumption which now obtains is so excessive that the aqueduct cannot convey sufficient water to satisfy it. New schemes for procuring additional supplies are under consideration, and measures for suppressing waste are receiving serious attention.

In Chicago the consumption, which in 1858 was 27·3, is now 102·2 gallons per head.

In Brooklyn, when the present supply was introduced, in 1859, the consumption was at the rate of 21 gallons per head. It has now risen to 52½; and that a large part of this is wasted has been proved by measuring the quantity flowing into the distributing mains from 1 to 3 a.m.

In Boston, the consumption per head per day was 23 gallons in 1849, and 50 gallons in 1870. It has now risen to 62. In 1873 the Boston water engineers made some observations to determine the rate of waste by measuring the water flowing out of the Beacon Hill Reservoir between midnight and 3 a.m., into a district containing about 60,000 inhabitants. At the beginning of the experiment "the consumption was found to be somewhat irregular, but between 1 and 3 o'clock it was remarkably uniform, showing that the draft was not due to irregular opening and shutting of cocks, but to a continuous flow at almost unvarying outlets,"—in other words, it was due to continuous waste. The quantity that passed into the district between 1 and 3 a.m. was 322,251 gallons, being equal to a rate of 64 gallons per head in 24 hours.

That all American cities are not equally extravagant in the consumption of water is shown by the city of Providence, where, notwithstanding that waterworks have been constructed which are capable of supplying about five times the quantity now delivered, the total daily supply is only 21 gallons per inhabitant. This includes an unusually large supply per head to manufacturers, and water is said to be liberally used for all purposes.

The consequence of neglect and wasteful habits are most seriously felt when available resources approach exhaustion, and the expenditure of a large amount for additional supplies has to be faced. By this time the number of defective appliances has become so great, and the interests affected by any proposal to condemn existing arrangements have become so numerous and powerful, that it is often found to be easier to obtain powers to carry out a scheme for a further supply, involving a large outlay, than it is to obtain, and put into operation, powers to interdict the use of wasteful fittings.

The difference between the first cost of providing and fixing proper and substantial pipes, cocks, and other apparatus, and the first cost of flimsy and wasteful fittings is insignificant, but to remove defective appliances and substitute approved fittings is often a very costly undertaking. The first cost of fittings adapted for a constant supply is less than for an intermittent supply.

Excessive waste of water is attended by no corresponding advantage. There is a popular impression that the cleansing of drains and sewers is promoted by the waste from taps and closets; this is a mistake. Drains and sewers can only be effectually flushed by the sudden discharge of a sufficient volume of water from their highest points, and the quantity required to do this is very small in proportion to the total consumption in any town.

A very large part of the needless waste that takes place percolates unto the foundations of houses, and often completely saturates the subsoil. In this way very great injury is done to the health of the inhabitants.

The consequence of neglecting to enforce proper regulations may thus be summed up:

1. Excessive waste.

2. Increased cost of procuring and distributing water, and consequently higher charges.

3. Premature exhaustion of waterworks capabilities.

4. Danger to health from allowing imperfect appliances and careless plumbing, and from dampness of subsoil.

5. Difficulty of getting inferior fittings removed after they have been once fixed.

SOURCES OF WASTE.

In describing the principal sources of waste, I shall, in keeping with the character of this book, confine my remarks almost entirely to waste which is directly under the control of consumers, and for which they are more or less responsible.

In places where the public mains have been in the ground for a long period, and especially where rival water companies have been contending for customers, or where a lax system of management has prevailed, there may be considerable waste from the mains and street service-pipes; but in the majority of towns, and as a general rule, the bulk of waste takes place from private pipes and fittings. It may be conveniently classified under the following heads:—

1. Waste from badly designed apparatus, defective workmanship, and the employment of inferior material.

2. Superficial waste due to wear and tear.

3. Hidden waste from broken pipes underground, and from sources not visible on a superficial examination.

4. Taps left open: this is an evil that attends the intermittent system, and seldom gives any trouble under the constant system.

PREVENTION OF WASTE.

Regulations for the prevention of waste necessarily vary according to the powers contained in the special Acts of waterworks authorities. The Waterworks Clauses Act, 1847, contains provisions under which pipes to be laid by consumers are required to be of approved strength and material; and where by the special Act the water need not be constantly laid on under pressure, every person supplied is required to provide a proper cistern ball and stop-tap, and to keep the same in good repair. The Act also imposes a penalty of £5 for wilful waste. The provision with respect to cisterns and taps does not apply to districts supplied under the constant system. The same Act enables waterworks officers to inspect premises in order to ascertain if there is any waste or misuse of water.

The Waterworks Clauses Act, 1863, contains further provisions for the protection of water: it gives power to cut off in certain cases, and imposes a penalty for causing or suffering waste, misuse, or contamination.

By the Metropolis Water Act, 1852, powers were granted to the London water companies (subject to the approval of the Board of Trade) to prescribe the size, nature, and strength of pipes, cocks, and other apparatus, and to interdict any arrangements and the use of any pipes, cocks, and other apparatus which might tend to waste or misuse. These powers were not exercised until after the passing of the 1871 Act.

In 1859 the Norwich Water Company, finding the waste so great that their works were insufficient to keep up the supply, and having failed to reduce it by the provisions of the Waterworks Clauses Act, applied to Parliament for, and obtained, absolute power to prescribe the size, nature, strength, materials, mode of arrangement, and repair of pipes, cocks, cisterns, and other apparatus; and to interdict any arrangement, or the use of any pipe, cock, &c., which in their judgment might tend to waste, misuse, undue consumption, or contamination. These powers were exercised without delay, and with great benefit alike to the company and to the consumers. The consumption was reduced from 40 gallons per head per day to 14½ gallons per head per day, for all purposes.

In 1860 the Manchester Corporation obtained from Parliament powers resembling those granted to the Norwich Company. Many other towns have since then succeeded in getting powers of the same kind, although not always as complete and absolute. In all cases where such provisions have been put into operation, they have produced excellent results in diminishing waste.

Some years ago, the Manchester Corporation introduced a method of controlling plumbers; and they subsequently established an office for testing and stamping fittings. No tap, closet, cistern, or other apparatus is allowed to be fixed, unless it bears the stamp of the Corporation Testing Officer. Any person may send fittings for examination on payment of a small fee. Under this system, the public have a guarantee for the quality of the articles they buy; and the water authorities prevent the employment of incompetent plumbers and the erection of inferior fittings. A superficial inspection, made after fittings have been fixed, is not sufficient to insure proper construction and quality of material.

In some towns the water authorities endeavour to secure sound fittings, by requiring them to be made by a specified manufacturer. This plan is objectionable because it gives a monopoly to one or a few makers, and subjects buyers to additional expense. The advantage is claimed for it that it secures uniformity and interchangeability in the parts; but this advantage can be otherwise attained.

Where the pipes, cocks, and other apparatus are all in conformity with the best regulations it is easy to maintain a constant supply with a low rate of consumption, but where no proper regulations have been enforced, and where the house fittings are consequently of an inferior class, the consumption is found to increase in proportion to the neglect that has prevailed. To effect an entire, or even a considerable, change in the house fittings of a town is no light undertaking. The cost of the alterations is felt by owners and occupiers to be a grievous burden, and it is not without reason that they complain when authorities interdict arrangements which have originally been made, if not with their express sanction, at least without any objection on their part.

An attempt was made in Liverpool, in 1874, to promote a bill containing provisions similar to those given to Norwich, Manchester, and other places. Owing to excessive waste, the Liverpool water supply was in a most critical state, and those who were responsible for its management were advised that the waste could only be effectually repressed, and the supply to the city maintained, by adopting measures of control, such as had been successful elsewhere. The City Council approved of a Fittings Bill, but its promotion was opposed, under the Borough Funds Act, by an association of houseowners. A poll was taken, and there was an overwhelming majority against proceeding with the measure. Consequently it had to be abandoned. Under these circumstances it became necessary to attack the waste with the limited powers which the corporation possessed. Out of this necessity a system was gradually evolved which produced unexpected results. As now fully developed the Liverpool system is briefly this:—

The area of supply is divided into 205 districts, containing an average population of about 3,100. The supply to each district is controlled by a waste water meter, which, as the water passes through, marks on a diagram the volume flowing at any and every instant. Every separate communication-pipe to a house, or block of houses, or other building, is controlled by a stopcock laid under the footway, as described at p. 125, and as is the practice in many towns. The meters indicate the consumption in the districts, and when the diagrams are examined, the bad can at once be separated from the good. The lines drawn on the diagram show when water is being legitimately used; the waste is indicated by a steady horizontal line. The difference in the amount of waste between two adjoining districts of equal population is often 200 or 300 per cent. Only the wasteful districts are selected for examination, and these are visited by a small staff of night inspectors, who begin their operations at about midnight. Between midnight and 5 a.m. most of the water flowing into a district is going to waste, and experience has shown that this waste is not distributed equally over all the houses in the district, but is confined to about 5 per cent. of them.* That is to say, given a district containing 500 houses in which the total rate of waste is 1,000 gallons an hour, the whole of that waste is to be found in 25 of the houses. Now the task which the inspectors are set to perform is to discover in which of the houses in a specific district the waste is taking place. This they accomplish by sounding the stop-cocks. When a night inspector proceeds to a district he opens the cover of the first stop-cock he reaches. He places one end of a steel bar on the top of the cock and listens at the other end. By practice he can hear almost any flow, however slight, that there may be through the cock. If there is no noise of water passing through, he goes on to the next stop-cock. When he finds one on which he hears a noise he shuts it and again listens. If the noise continues after the cock is closed

* These figures apply to the present state of the fittings. During the transition period the proportion was fully 25 per cent.

it evidently comes from the street (or main) side, but if the noise has ceased, it is proved to have been due to water flowing into the premises. Any cessation of flow that may be caused by shutting a stop-cock affects the district meter, and is recorded on the meter diagram. Before a night inspector leaves a district he re-opens all the stop-cocks which he has closed for experiment. His discoveries are reported at the office, and on the following morning a day inspector visits the houses supplied from the stop-cocks reported, to search for the waste. The day inspector knows that there is a certain quantity of waste to be accounted for, and if he cannot find it by a superficial examination, he will probably find it underground. Any one can appreciate the difference between sending a police officer to look for a criminal in Smith Street on a probability of finding him there, and sending to a specific house in Smith Street on positive information that he was there on the previous night.

The essence of the Liverpool system consists in the localization of waste, first to a district by the meter records, and then to certain houses or fixed points in the district by night inspections. Perhaps the most remarkable result achieved by the system has been in disclosing the existence of hidden underground waste, of which no trace or sign reaches the surface. This kind of waste takes place to an extent of which few persons have any conception. The Liverpool communication-pipes are certainly not inferior in quality or strength—they are probably superior in strength—to the average of lead water-pipes. Yet, out of every 1,000 leakages from iron and lead pipes traced by the waste water inspectors, 568 are underground leakages (frequently of a very serious character) which do not rise to the surface.

The success of the Liverpool experiment has been complete and gratifying. Under an intermittent supply of 9½ hours' duration per day, the average consumption was over 29 gallons per head per day, and when constant service was experimentally turned on the consumption rose to over 37 gallons per head. An unrestricted constant service is now given with an average consumption of 22½ gallons per head per day for all purposes. The difference between the former intermittent supply and the reduced constant supply amounts in money value, at the cost price of the water, to £53,000 per annum.

This change has been effected, and these results have been achieved, without any sweeping condemnation of faulty appliances, and therefore without the application of the rigorous powers which have enabled similar results to be accomplished in other places. The waste water meter system is now in operation in several towns.

In Liverpool the water inspectors are practical plumbers,

and they repair free of charge all simple defects, such as regulating wires, valves, or levers, and renewing washers. For this purpose they carry a small bag of tools. Landlords and tenants are saved from plumbers' bills for trifling repairs; the waste is immediately stopped; and the labour and expense of issuing and serving notices and making re-examinations are avoided. I understand that the same practice has lately been commenced in Hull, with results equally satisfactory.

The consequences of permitting excessive waste cannot be too strongly urged upon the attention of owners and occupiers of property, so that they may see how deeply they are interested in the maintenance of an economic system of distribution, and how greatly it is to their advantage to co-operate with water authorities in carrying out proper regulations for waste prevention. If inferior workmanship and imperfect house fittings are allowed, they produce evils from which all must suffer.

The precautions required against waste are simple, and not expensive. There is no excuse for neglecting them. The regulations to be observed in regard to fittings are chiefly these:—The pipes to be of material and strength suited to the ground in which they are to be laid and to the water they are to convey. Taps to be of the screw-down kind, strong, well made, and of the best hard brass. Water-closet apparatus to give an efficient flush, but not to allow a continuous discharge. All overflow and waste-pipes to be brought to the outside of premises, to conspicuous points.

Questions often arise as to who is responsible for the repair of house fittings. The statutory powers under which water authorities issue notices requiring defects to be repaired generally apply to the "persons supplied with water," and it is these persons, whether owners or occupiers, who are liable for the penalties imposed for waste or misuse. At the same time it is the custom in many towns for owners of property to execute all repairs that are required, though they are not compelled to do so, in the absence of an agreement, if the occupier pays the water rates.

CHAPTER IX.

RURAL SUPPLIES.

General state of water supply in rural districts—Improvements needed—Polluted streams and wells—Examples—Shallow wells—How to provide wholesome water—The Abyssinian tube—Rain-water from roofs—Filtration—Machinery for raising water.

ACCORDING to the sixth Report of the Rivers Pollution Commissioners about "twelve millions of country population (in Great Britain) derive their water almost exclusively from shallow wells, and these are, so far as our experience extends, almost always horribly polluted by sewage and by animal matters of the most disgusting origin." Though this may be, and doubtless is, an accurate description of the shallow well samples analysed by the Commissioners, it is certainly an exaggerated picture of the general condition of our rural water supply. At the same time no one who has investigated the subject will question the statement that the domestic supply of the country population is characterized by deficiency in quantity and inferiority in quality, and that there has long been a pressing call for a vigorous measure of reform.

The Public Health Amendment (Water) Act, 1878, contains provisions by means of which, if they are resolutely applied, a great improvement can be speedily effected without any undue or vexatious interference with the carrying on and full development of agricultural, mining, and manufacturing industries. The leading principle of the Act is that the erection or ownership of a dwelling-house carries with it an obligation to provide for the inhabitants a wholesome supply of water available within a reasonable distance. So far as new houses are concerned the Act leaves little to be desired, for no house can in future be occupied without an official certificate having been obtained declaring that a satisfactory supply has been provided. In regard to existing houses the provisions are necessarily less stringent. Their success will depend upon the energy and ability of district health officers, and the conscientiousness of rural sanitary authorities.

There are scores, and probably hundreds, of thousands of country houses whose polluted water supply might be made wholesome by very simple and inexpensive measures, which could be carried out in a few months if the rural sanitary authorities were to set themselves earnestly to accomplish it. The pollution is caused by sheer stupidity and slovenliness in arrangements for disposing of house sewage and farmyard manure. Privies and manure-pits are placed in such positions that their contents contaminate wells, and are washed into streams instead of being employed to fertilize the land. It is very common to find watercourses flowing through, or in close proximity to, farmyards. Into these watercourses the house slops are conveyed, sometimes by a drain-pipe, sometimes in an open channel. Whenever a heavy fall of rain takes place the filth about the farmyard, from the house, stables, shippon, and piggeries, is swept into the brook. Nothing but the deplorable ignorance of the offenders can be urged in palliation of the recklessness with which streams are fouled in utter disregard of the comfort and health of subsequent users. Every nuisance of this kind ought to be immediately abated, and the offenders compelled to store and utilize their sewage so as to cause no injury to their neighbours.

The illustrations that follow represent two cases that have recently come under my notice, and I believe they cover a large class.

In the first case an excellent mountain stream affords a supply, for all domestic and general purposes, to a country house, and, at a lower point, is used for the supply of a small village. The building overhanging the stream is a servants' privy. Special provision had been made for preserving the purity of the water by collecting all the sewage from the hall and outbuildings in cesspools, but it was found inconvenient to have the

sewage put on to the meadow during hay-time, and consequently the water-closet soil-pipe and the kitchen drains were turned into the stream, in the manner indicated by the dotted lines. The connection was of course not so apparent on a superficial examination as is shown in the sketch.

The next case illustrated is that of a farmer polluting the water supply of his neighbour.

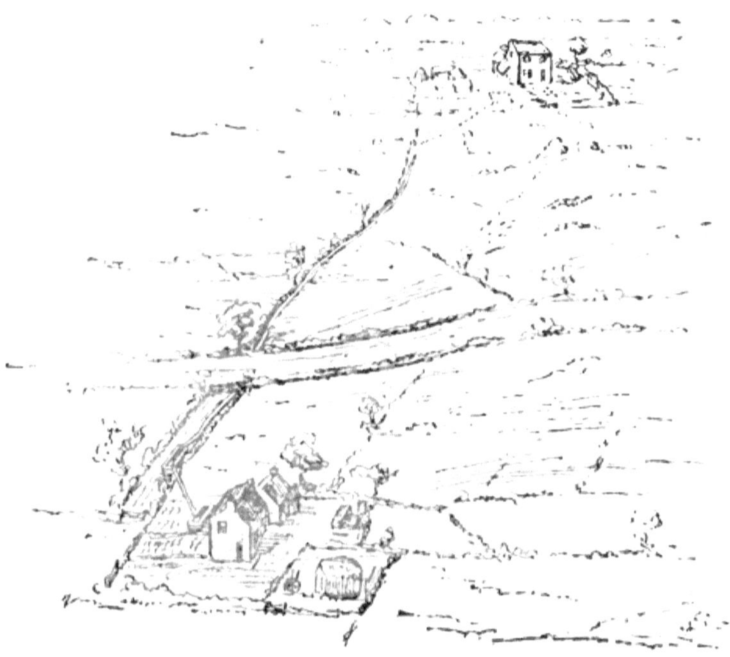

The lower farm has, at the road-side, a tank for cattle to drink from, and, nearer the house, a tank for domestic purposes; these are both fed from a stream seen descending through the upper farm. A short time ago the cattle refused to drink from their tank, and a foul smell was perceived from the house tank. On investigation it transpired that on farm No 1 a new arrangement had been made for collecting the sewage from the farm and some adjoining cottages into a cesspit, from which an overflow (shown by the dotted line) had been laid, underground, to the watercourse; and as the cesspit was rarely emptied the brook became the ordinary channel for the sewage.

In the majority of instances an experienced observer has no difficulty in pointing out sources of pollution. But it frequently happens that places which on a superficial inspection appear to be clean, well arranged, and in every respect satisfactory, are found, on a closer examination, to belong to the worst class of cases. The filth is carried off in underground channels, which are laid no one knows how, and which end no one can tell where. That they terminate in a watercourse may safely be assumed.

The ordinary shallow wells of country places are too well known to need describing. There are many of them that yield safe drinking-water, but the majority are undoubtedly contaminated, more or less, by infiltration from cesspools, by filth from the surrounding surface, or by drainage from manured lands. The existence of a well from which a household is regularly supplied may generally be regarded as evidence that good well-water is to be had in the neighbourhood if only proper precautions are taken against contamination. To line the sides of the well with brick or stonework down to the rock, as is commonly advised, and as ought always to be done, is of little avail if there are cesspools about. There are many villages like Bourton, of which Dr. Ballard, a medical officer of the Local Government Board, reported a few years ago that it was "riddled with cesspits. As soon as one cesspit is full, instead of emptying it, they simply leave it alone and dig another." Under such conditions the well-waters can only be made fit to drink by carrying out a proper system of sewerage for the district. When this has been done all the cesspits must be carefully emptied of filth and filled with clean soil.

In lining a well to prevent the entrance of filth near to the surface, it is necessary not only to build a substantial wall, but also to give it a backing of concrete, or of clay worked into puddle, in a dense, putty-like condition, and carefully put in so as to form a compact wall; or to line the well with an iron cylinder. The lining should be firmly built in to the water-bearing stratum. It is always desirable to carry the top of the well a few inches above the ground level, and to provide a cover.

If a cesspool is indispensable it should be made perfectly water-tight, and as far as possible from any well, and no overflow should be allowed.

The following illustration is copied from the Tenth Annual Report of the Massachusetts State Board of

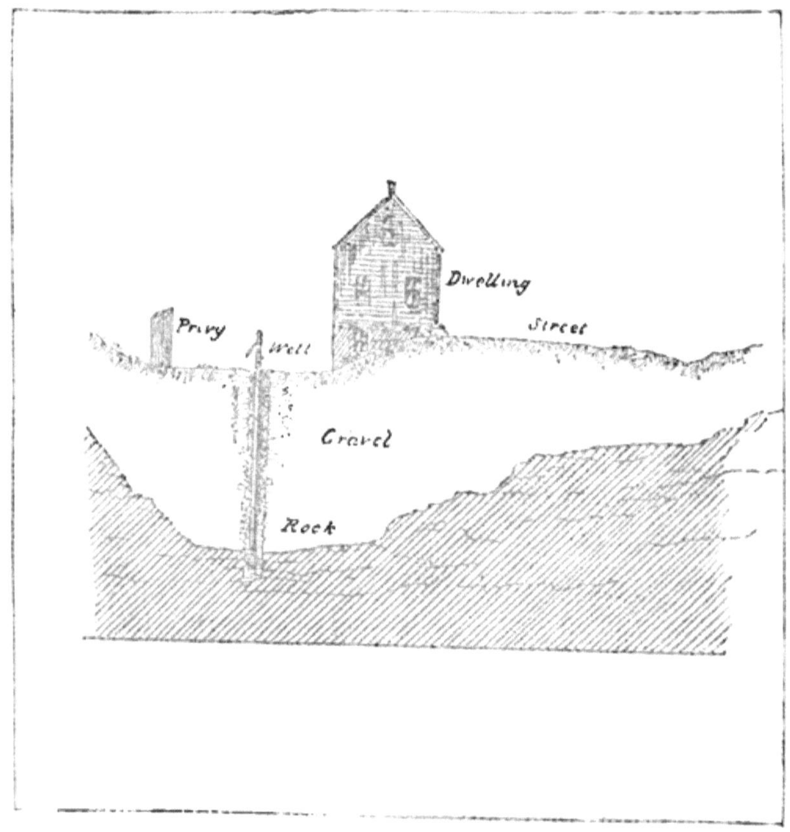

Health. I reproduce it here as representing a class of cases common enough in this country, as well as in the United States; and because the persons using the water

would, if they thought about the matter at all, probably not suspect any evil effects from it.

There were twelve cases of typhoid fever among persons using this well-water; the house became the centre of infection for a whole neighbourhood.

The measures required to provide a wholesome supply for country houses are generally of a simple, and frequently of an inexpensive, character. In hilly districts, springs abound; and where they can be made available they give water of the finest quality. Surface-water from pasture and meadow-lands may safely be used, if sewage is excluded. Water from arable-lands is not suitable for drinking, without efficient filtration. In manufacturing districts, streams are seldom sufficiently free from pollution to be used for drinking; but all our important manufacturing districts are on geological formations, from which water can be procured by wells.

Where the conditions of the ground are suitable, and where water can be obtained within 30 feet from the surface, the Abyssinian well-tube is a simple and cheap combination of well and pump, which possesses many advantages over the ordinary form of well. The Abyssinian tube derives its name, I believe, from its successful employment for supplying the British army during the expedition into Abyssina. It is merely a tube, with perforations at one end, and a steel point, driven into the ground, with a pump attached to the top of the tube.

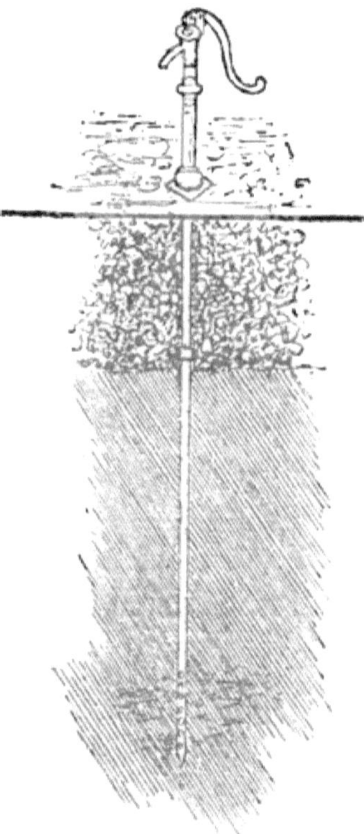

If no water is found in one place, the tube can be withdrawn, and driven elsewhere. In this manner the water yielding capabilities of a locality may be easily tested.

If wholesome water cannot be obtained from springs, wells, or streams, recourse can always be had to the collection of rain from the roofs of buildings, in tubs or tanks. For this purpose, the roof must be constructed of material that will not communicate any impurity to the rain falling upon it. Thatched or felt roofs are obviously unsuitable.

Then, the roofs must be fitted with gutters and downspouts for collecting and conveying the rain. There are in England many thousands of farmhouses and cottages that are not provided with spouting.

Rain-water-tanks are usually built underground. It

is found to be more convenient and on the whole cheaper to construct the tanks underground than on the surface, and the water is kept at a more uniform temperature. The objections to underground tanks are that they often leak without the defect becoming known until they run dry; that they are liable to pollution; and that a pump is required to draw water if due regard is to be had to comfort and cleanliness. Tanks formed aboveground have the advantage of being more easily examined, defects are more readily disclosed, they are free from the risk of surface-pollution, and the water can be drawn by opening a tap or pulling out a plug. The best shape for a tank is circular, or rectangular with the corners rounded off. The material may be the ordinary building material of the locality. The object to be aimed at is to secure a strong, water-tight, and durable receptacle, and any material that will fulfil these conditions may be employed. Portland cement concrete (six parts of gravel and broken stone to one part of the best cement) is an excellent material for the purpose, but to be properly made it requires more experience in mixing than ordinary country stonemasons and bricklayers possess.

Every tank should be well covered, and a manhole provided, protected by a flag or by timber. Care must be taken that there is no bright lead or zinc about the roof or cistern. As explained elsewhere, exposed lead is soon tarnished, and in that state is not affected by soft water. To determine the tank capacity required for a household depending entirely on stored rain-water for their supply there are four elements that enter into the calculation.

1. The amount of rainfall.
2. The area of the collecting surface.
3. The period of longest drought.
4. The number of gallons per day to be provided for.

I shall not ask the reader to follow this calculation through for various parts of the country, but the practical result is that a tank for an ordinary cottage supply should be capable of holding from 1,000 gallons on the west coast to 1,500 gallons on the east coast; and this is on the assumption that the collecting surface is large enough to give the average daily supply required with the minimum annual rainfall. To ascertain approximately the average number of gallons per day that can be had from a roof, with adequate storage space, multiply the lowest recorded rainfall in feet, by the collecting area in feet, by ·015. This factor includes a small allowance for loss in the collection. A few data with respect to water may be useful here.

1 cubic foot of water is equal to 6·23, say 6¼, gallons.			
1 gallon	,,	contains	277·27 cubic inches.
1 ,,	,,	weighs	10 lbs. avoirdupois.
1 cubic foot	,,	,,	62·35 lbs. ,,

These figures refer to pure water at a temperature of 62 degrees F., with the barometer at 30 inches.

One inch of rain over 100 square feet of surface amounts to 52 gallons. In measuring the area of a roof the slopes must not be taken into account. Only the horizontal area covered by the roof must be measured.

The average rainfall in Great Britain varies from about 20 inches on the east coast to 70 or 80 on the mountain districts of Wales, and on the watersheds of the Westmoreland and Cumberland lakes. Exceptional falls of 200 inches have been recorded in the Lake district.

For farms and large country residences where more water is required than can be collected from the roofs it may be necessary to construct a small artificial gathering ground, but additional collecting area can in most instances be provided in conjunction with other improvements, such as permanent rick-covers, cattle shelters, or glass houses for early market produce.

Rain-water cannot be collected from roofs in a tolerably clean state if attention is not paid to the spouting, to prevent birds building their nests, and otherwise fouling the water-channels. Even with the best attention, the water will be contaminated by dust and soot, and by the droppings of birds. Though water collected in this manner is largely used for potable purposes without any filtration, it is seldom that the conditions are so favourable as to afford a satisfactory supply throughout the year. Filtration of some kind has generally been deemed necessary, and many plans have been proposed and tried. The most common is the placing of filtering material (sand, gravel, and charcoal) at one end of the tank, in a compartment separated from the other part of the tank by a partition open at the bottom. Another method is to make the water pass through a porous brick partition, in plan thus:—

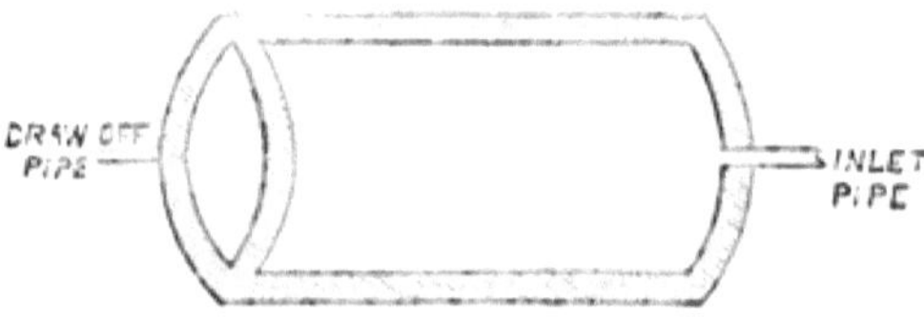

When the pores of the bricks become clogged, as they do in course of time, they must be brushed, or the partition renewed.

Another plan is to put the end of the pump suction pipe in a small filter, which can be lifted out of the tank for cleaning. A combined tank and filter is shown at page 56, but all arrangements under which it is necessary to empty the tanks before the filtering materials can be examined, washed, or renewed, are objectionable.

On the whole, I think it better not to attempt any purification of the water in the tanks, unless it be a simple straining to keep back the grosser impurities, which can be effected by a wire gauze strainer fitted into a frame, sliding in a groove, and easily drawn out. Water for drinking can be more effectually filtered in the house.

In a case of which I have recently been informed, a complaint that the tank-water was unsatisfactory was remedied, without filtration, by building a partition across one end, to allow the impurities to deposit, in this way:—

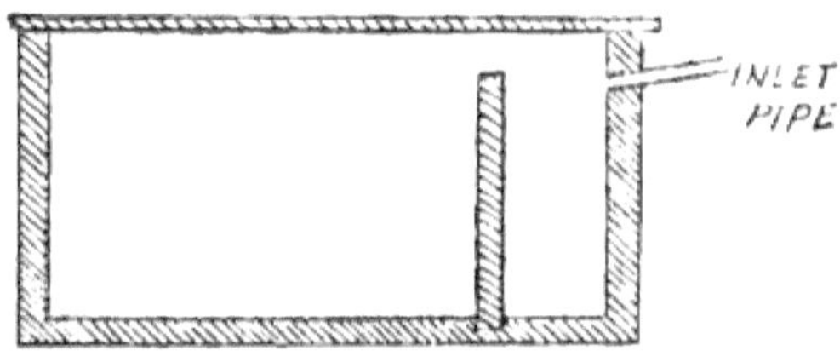

The water has, of course, to pass over the partition from the subsiding compartment into the main tank. The inlet should be arranged so that the water will enter with as little disturbance of the tank contents as possible.

For raising water by machinery to supply villages, farms, or large residences, the hydraulic ram is the most simple and durable, and probably the cheapest apparatus, that can be employed. It is now made so that it can be worked by impure water, and a polluted stream can be utilized to lift water from a well, or to throw surface water to any desired elevation. The only working parts liable to derangment are the valves, which must be occasionally examined and renewed.

The turbine, or water-wheel, can also be driven by foul water, and requires but little supervision: the pumps and gearing are, however, sometimes troublesome. In the Fen districts, wind-engines are occasionally to be met with, and in other countries they are more common than in England: where water power cannot be had, they may be used with advantage. The necessity of furnishing gearing for working them by horses or by steam, in the absence of wind, adds to their cost. In places where there is a supply of gas at a moderate price, a gas-engine will be found an economical motor: it can be managed by a labourer of ordinary intelligence.

CHAPTER X.

WATER FOR TRADE PURPOSES.

Distinction between domestic and trade supplies—Practice as to charging by meter—Directions for reading meter index—Precautions against waste and overcharge—Examples of rates charged in various towns—Water as motive power—Advantages of water-pressure machinery.

WATERWORKS proprietors are not legally bound to supply water for trade purposes, as they are, on certain conditions, bound to supply water for domestic purposes. The maximum rates that can be charged for domestic supplies are generally fixed by Act of Parliament, but charges for trade supplies are left to be the subject of agreement between the waterworks authority, and the person requiring the supply. A definition of what is included in a domestic supply is usually given in the special Act of every water authority. In the Waterworks Clauses Act, 1863, it is enacted that—"A supply of water for domestic purposes shall not include a supply of water for cattle, or for horses, or for washing carriages where such horses or carriages are kept for sale or hire, or by a common carrier, or a supply for any trade, manufacture, or business, or for watering gardens, or for fountains, or for any ornamental purpose."

This enactment only applies to special waterworks Acts passed since 1863, and incorporating the Clauses Act; but a similar clause is to be found in many earlier private Acts.

In the Metropolitan Water Companies' Acts, a domestic supply is defined thus:—"That a supply of water for domestic purposes shall not include a supply for steam-engines or railway purposes, or for warming or ventilating purposes, or for working any machine or apparatus, or for baths, horses, cattle, or for washing carriages, or for gardens, fountains, or ornamental purposes, or for flushing sewers or drains, or for any trade, or manufacture, or business requiring an extra supply of water."

The only exception to this is the East London Company's Act, which differs only in that it includes under domestic purposes a supply for baths in houses of which the annual value exceeds £30.

The usual rule in towns is to supply all trade purposes requiring a considerable or uncertain quantity of water, through meter; but for small trade purposes, such as for building, or small steam-engines, or for washing carriages, for cattle, and watering gardens, it is usual to make an annual charge by assessment. The meters for measuring water are, in most places, provided by the waterworks authorities, who charge a sum as rent, which covers interest on the price of the apparatus, and the expense of inspecting and repairing it. This amounts to about 10 per cent. per annum on the first cost of the meter.

Every meter has a visible index attached to it, and consumers will find it useful to keep a daily record of the quantity indicated on the dial. Such a record is of service as a check upon the consumption, and calls immediate attention to waste or misuse. It is also a check upon the accuracy of the water bill.

The following represents an index dial, with an example of a reading.

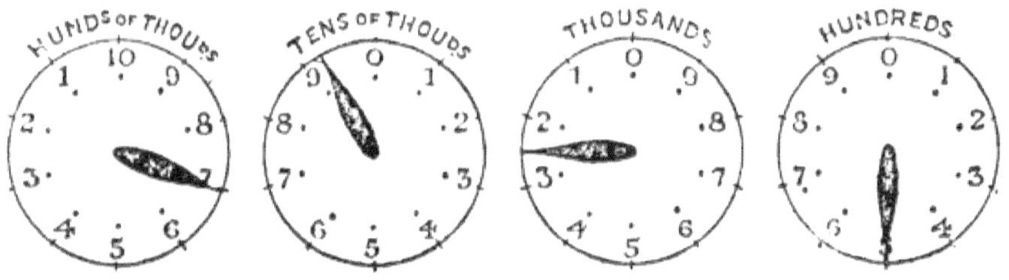

The quantity indicated in this example is 692,500 gallons. Some dials extend to tens, and the dial shown, though it is only marked for hundreds, can be read to tens by observing the position of the pointer between the figures on the hundreds circle.

If a consumer has reason to doubt the accuracy of a meter, the water authorities will generally, upon application, test the instrument at their works, on condition that the applicant undertakes to pay the expense of the removal and test if the meter proves to be registering correctly the quantity passing through it. Trade consumers frequently complain of being debited with more water than they believe to have been consumed, but upon investigation such complaints are almost invariably traced to waste from defective fittings, or to undue consumption, of which the persons complaining have been ignorant. Hence the importance of adopting the suggestion made above, to enter in a book, at a fixed hour every day, the reading of the meter index.

Leakages from underground pipes on the outlet side of meters may be readily detected by closing all the draw-off taps and noticing if water is still passing through the meter. Trade consumers will find it an economical practice to shut the stop-tap at the meter whenever water is not wanted.

An useful check upon the operations of a manufacturing establishment may be provided by attaching to the meter a clock and diagram which will show the precise moment at which the workmen begin to use or cease using water, and the quantity passing through the meter during any part of the day.

The rates charged for water supplied through meter are, in the majority of towns, upon a sliding scale under which large consumers get water at a lower price than small consumers. In some towns a fixed charge per thousand gallons is made to large and small consumers alike. The sliding scales that are adopted in various towns differ greatly in their range and design. In the following table the minimum and maximum rate per 1,000 gallons is given in each case where the charges are by a sliding scale. In comparing the figures it must be remembered that hardly any two scales begin and end at the same points. (See Table, p. 176.)

WATER AS MOTIVE POWER.

Where water is cheap, the supply constant, and delivered under a considerable pressure, it may, with great advantage and economy, be used for many purposes of power instead of steam or manual labour. There are many towns in England and in foreign countries where water from the ordinary distributing mains is extensively used as a motive power. In Liverpool there are about eighty lifts and hoists, and a number of organs, worked by water-pressure from the corporation mains. In Zurich, Swit-

EXAMPLES

OF

RATES CHARGED PER 1,000 GALLONS FOR TRADE SUPPLIES BY METER.

	Maximum.		Minimum.		
	s.	d.	s.	d.	
Aberdeen	0	8¼	0	5¼	
Ashton-under-Lyne	1	0	0	6	
Birkenhead	1	0	1	0	
Birmingham	2	0	0	7	
Blackburn	2	0	0	6	
Bolton { in the borough	0	6	0	6	
Bolton { outside „	0	9	0	9	
Bradford	1	0	0	3½	
Brighton	1	3	0	9	
Bury	2	0	0	7½	
Carlisle	0	10	0	6	
Dundee	0	7	0	7	
Edinburgh	0	9	0	9	
Glasgow	0	4	0	4	
Halifax	0	10	0	6	Beyond borough 50 per cent. extra
Huddersfield	1	6	1	0	
Hull	0	9	0	6	
Liverpool { in the borough	0	7	0	7	
Liverpool { outside „	0	9	0	9	
London—					High service 25 per cent. additional
East London Co.	0	9	0	6	
Grand Junction Co.	0	9	0	6	Ditto
New River Co.	0	7½	0	6	
Southwark and Vauxhall Co.	0	9	0	6	Ditto
West Middlesex Co.	0	9	0	6	Ditto
Chelsea Co.	1	0	0	6	
Lambeth Co.	1	0	0	6	
Manchester	2	0	0	6	
Nottingham	1	0	0	6	
Perth	0	9	0	6	
Plymouth	0	2	0	2	
Preston { in the borough	1	0	0	4½	
Preston { outside „	1	6½	0	7	
Sheffield	1	0½	0	6	
St. Helen's	0	5¼	0	5¼	
Wolverhampton	1	4	0	6	
Worcester { city	0	5	0	5	
Worcester { outside	1	0	1	0	
York	1	6	0	6	

zerland, a town of about 20,000 inhabitants, there are 113 water-engines, varying in size from one-third to 4 H.P., at work for various trade purposes. The average pressure from the mains is about 50 lbs. per square inch, and the charge for water is at the rate of 5*d.* per indicated horse-power per hour.

Among the advantages which may be claimed for water-pressure machinery are:—

1. Steadiness, ease, precision, and comparative noiselessness of action.

2. Absence of risk from accident.

3. Facility with which the movement of a machine may be controlled from any point of its travel.

4. Simplicity of working, rendering the employment of skilled labour unnecessary.

5. Saving in charges for insurance, as compared with machinery involving the use of fire or light.

6. Limitation of the expenditure of power to the time during which useful work is being performed, thus enabling machinery to be employed intermittently without loss of power.

7. Ease with which energy may be transmitted to considerable distances without appreciable loss.

8. The opportunity afforded of making special provision for extinguishing fires by attaching fire hydrants to the pipes laid to convey the water-pressure.

CHAPTER XI.

REGULATIONS FOR THE PREVENTION OF WASTE AND MISUSE OF WATER.

As far back as 1852 the metropolitan water companies obtained statutory powers to make regulations for preventing waste and misuse of water, subject to the approval of the Board of Trade. As these powers were not exercised, Parliament, in the Metropolis Water Act of 1871, made it compulsory on the part of the companies to make such regulations within six months after the passing of that Act.

The companies accordingly prepared regulations which were submitted to three Commissioners (Lord Methuen, Captain Tyler, and Mr. Rawlinson), on behalf of the Board of Trade. An exhaustive inquiry was held, at which the water companies, the Metropolitan Board of Works, and the City Council were represented. The regulations which were settled by the Board of Trade after this inquiry have been circulated by the Local Government Board as specimen regulations for the information and guidance of other towns.

As these regulations apply to all the inhabitants of the metropolis, and embody most of the provisions usually inserted in recent waterworks regulations, they are set forth here.

REGULATIONS MADE UNDER THE METROPOLIS WATER ACT, 1871.

1. No "communication-pipe" for the conveyance of water from the waterworks of the Company into any premises shall hereafter be laid until after the point or place at which such "communication-pipe" is proposed to be brought into such premises shall have had the approval of the Company. Place of communication-pipe.

2. No lead pipe shall hereafter be laid or fixed in or about any premises for the conveyance of or in connection with the water supplied by the Company (except when and as otherwise authorized by these regulations, or by the Company), unless the same shall be of equal thickness throughout, and of at least the weight following, that is to say:— Weight of lead pipes.

Internal Diameter of Pipe in Inches.	Weight of Pipe in lbs. per lineal Yard.
⅜-inch diameter.	5 lbs. per lineal yard.
½ " "	6 " "
⅝ " "	7½ " "
¾ " "	9 " "
1 " "	12 " "
1¼ " "	16 " "

3 Every pipe hereafter laid or fixed in the interior of any dwelling-house for the conveyance of or in connection with the water of the Company, must, unless with the Interior pipes.

consent of the Company, if in contact with the ground, be of lead, but may otherwise be of lead, copper, or wrought iron, at the option of the consumer.

Not more than one communication-pipe to each house.

4. No house shall, unless with the permission of the Company in writing, be hereafter fitted with more than one "communication-pipe."

Every house with certain exceptions to have its own communication-pipe.

5. Every house supplied with water by the Company (except in cases of stand-pipes) shall have its own separate "communication-pipe." Provided that, as far as is consistent with the special Acts of the Company, in the case of a group or block of houses, the water rates of which are paid by one owner, the said owner may, at his option, have one sufficient "communication-pipe" for such group or block.

No house to have connection with fittings of adjoining house.

6. No house supplied with water by the Company shall have any connection with the pipes or other fittings of any other premises, except in the case of groups or blocks of houses, referred to in the preceding Regulation.

Connection to be by ferrule or stop-cock.

7. The connection of every "communication-pipe" with any pipe of the Company shall hereafter be made by means of a sound and suitable brass screwed ferrule or stop-cock with union, and such ferrule or stop-cock shall be so made as to have a clear area of waterway equal to that of a half-inch pipe. The connection of every "communication-pipe" with the pipes of the Company shall be made by the Company's workmen, and the Company shall be paid in advance the reasonable costs and charges of and incident to the making of such connection.

Material and joints of external pipes.

8. Every "communication-pipe" and every pipe external to the house and through the external walls thereof, hereafter respectively laid or fixed, in connection with the water of the Company, shall be of lead, and every joint thereof shall be of the kind called a "plumbing" or "wiped" joint.

No pipe to be laid through drains, &c.

9. No pipe shall be used for the conveyance of or in connection with water supplied by the Company, which is laid or fixed through, in, or into any drain, ashpit, sink, or

manure-hole, or through, in, or into any place where the water conveyed through such pipe may be liable to become fouled, except where such drain, ashpit, sink, or manure-hole, or other such place, shall be in the unavoidable course of such pipe, and then in every such case such pipe shall be passed through an exterior cast-iron pipe or jacket of sufficient length and strength, and of such construction, as to afford due protection to the water-pipe.

10. Every pipe hereafter laid for the conveyance of or in connection with water supplied by the Company, shall, when laid in open ground, be laid at least two feet six inches below the surface, and shall in every exposed situation be properly protected against the effects of frost. Depth of pipes underground.

11. No pipe for the conveyance of or in connection with water supplied by the Company, shall communicate with any cistern, butt, or other receptacle used or intended to be used for rain-water. No connection with rain-water receptacle.

12. Every "communication-pipe" for the conveyance of water to be supplied by the Company into any premises shall have at or near its point of entrance into such premises, and if desired by the consumer within such premises, a sound and suitable stop-valve of the screw-down kind, with an area of waterway not less than that of a half-inch pipe, and not greater than that of the "communication-pipe," the size of the valve within these limits being at the option of the consumer. Stop valve.

If placed in the ground such "stop-valve" shall be protected by a proper cover and "guard-box."

13. Every cistern used in connection with the water supplied by the Company shall be made and at all times maintained water-tight, and be properly covered and placed in such a position that it may be inspected and cleansed. Every such existing cistern, if not already provided with an efficient "ball-tap," and every such future cistern, shall be provided with a sound and suitable "ball-tap," of the valve kind, for the inlet of water. Character of cisterns and ball taps.

Waste-pipes to be removed or converted into warning-pipes.

14. No overflow or waste-pipe other than a "warning-pipe," shall be attached to any cistern supplied with water by the Company, and every such overflow or waste-pipe existing at the time when these regulations come into operation shall be removed, or at the option of the consumer shall be converted into an efficient "warning-pipe," within two calendar months next after the Company shall have given to the occupier of, or left at the premises in which such cistern is situate, a notice in writing requiring such alteration to be made.

Arrangement of warning-pipes.

15. Every "warning-pipe" shall be placed in such a situation as will admit of the discharge of the water from such "warning-pipe" being readily ascertained by the officers of the Company. And the position of such "warning-pipe" shall not be changed without previous notice to and approval by the Company.

Buried cisterns prohibited.

16. No cistern buried or excavated in the ground shall be used for the storage or reception of water supplied by the Company, unless the use of such cistern shall be allowed in writing by the Company.

Butts prohibited.

17. No wooden receptacle without a proper metallic lining shall be hereafter brought into use for the storage of any water supplied by the Company.

Ordinary draw-tap.

18. No draw-tap shall in future be fixed unless the same shall be sound and suitable and of the "screw-down" kind.

Draw-taps in connection with stand-pipes.

19. Every draw-tap in connection with any "stand-pipe" or other apparatus outside any dwelling-house in a court or other public place, to supply any group or number of such dwelling-houses, shall be sound and suitable and of the "waste-preventer" kind, and be protected as far as possible from injury by frost, theft, or mischief.

Boilers, water-closets, and urinals to have cisterns,

20. Every boiler, urinal, and water-closet, in which water supplied by the Company is used (other than water-closets in which hand-flushing is employed), shall within three months after these Regulations come into operation, be

served only through a cistern or service-box and without a stool-cock, and there shall be no direct communication from the pipes of the Company to any boiler, urinal, or water-closet.

21. Every water-closet cistern or water-closet service-box hereafter fitted or fixed in which water supplied by the Company is to be used, shall have an efficient waste-preventing apparatus, so constructed as not to be capable of discharging more than two gallons of water at each flush. Water-closet apparatus.

22. Every urinal cistern in which water supplied by the Company is used other than public urinal cisterns, or cisterns having attached to them a self-closing apparatus, shall have an efficient waste-preventing apparatus, so constructed as not to be capable of discharging more than one gallon of water at each flush. Urinal-cistern apparatus.

23. Every "down-pipe" hereafter fixed for the discharge of water into the pan or basin of any water-closet shall have an internal diameter of not less than one inch and a quarter, and if of lead shall weigh not less than nine pounds to every lineal yard. Water-closet down-pipes.

24. No pipe by which water is supplied by the Company to any water-closet shall communicate with any part of such water-closet, or with any apparatus connected therewith, except the service-cistern thereof. Pipes supplyg. w.-closet to communicate with cistern only.

25. No bath supplied with water by the Company shall have any overflow waste-pipe, except it be so arranged as to act as a "warning-pipe." Bath to be without overflow pipe

26. In every bath hereafter fitted or fixed the outlet shall be distinct from, and unconnected with, the inlet or inlets; and the inlet or inlets must be placed so that the orifice or orifices shall be above the highest water-level of the bath. The outlet of every such bath shall be provided with a perfectly water-tight plug, valve, or cock. Bath apparatus.

27. No alteration shall be made in any fittings in connection with the supply of water by the Company without two days' previous notice in writing to the Company. Alteration of fittings.

Water-way of fittings.

28. Except with the written consent of the consumer, no cock, ferrule, joint, union, valve, or other fitting, in the course of any "communication-pipe," shall have a waterway of less area than that of the "communication-pipe," so that the waterway from the water in the district pipe or other supply-pipe of the Company up to and through the stop-valve prescribed by Regulation No. 12, shall not in any part be of less area than that of the "communication-pipe" itself, which pipe shall not be of less than a half-inch bore in all its course.

Weight of lead pipes having open ends.

29. All lead "warning-pipes" and other lead pipes of which the ends are open, so that such pipes cannot remain charged with water, may be of the following minimum weights, that is to say:—

½-inch (internal diameter)	- -	3 lbs. per yard.	
¾ "	do.	- -	5 " "
1 "	do.	- -	7 " "

Definition of "communication-pipe."

30. In these Regulations the term "communication-pipe" shall mean the pipe which extends from the district pipe or other supply-pipe of the Company up to the "stop-valve" prescribed in the Regulation No. 12.

Penalties.

31. Every person who shall wilfully violate, refuse, or neglect to comply with, or shall wilfully do or cause to be done any act, matter, or thing, in contravention of these Regulations, or any part thereof, shall, for every such offence, be liable to a penalty in a sum not exceeding £5.

Authorized officer may act for Company.

32. Where under the foregoing Regulations any act is required or authorized to be done by the Company, the same may be done on behalf of the Company by an authorized officer or servant of the Company, and where under such Regulations any notice is required to be given by the Company, the same shall be sufficiently authenticated if it be signed by an authorized officer or servant of the Company.

Existing fittings.

33. All existing fittings, which shall be sound and efficient, and are not required to be removed or altered under these Regulations, shall be deemed to be prescribed fittings under the "Metropolis Water Act, 1871."

www.ingramcontent.com/pod-product-compliance
Lightning Source LLC
Chambersburg PA
CBHW032143160426
43197CB00008B/766

* 9 7 8 3 3 3 7 1 3 9 3 4 6 *